A Glimpse at the Trave᷄ Baghdad

The history of Baghdad in the 18th and 19th centuries had predominantly been written by two groups. The first group is Baghdadi scholars, and the second group is travellers. These two resources complement each other; while the literature of Baghdadi scholars provides insights from inside, travelogues provide observations from outside. By implementing this interlocking method of investigation, we can reach a comprehensive understanding of the history of Baghdad. Having investigated some sources from inside in my previous book; *Baghdad: an urban history through the lens of literature*, the focus of this book is on travel literature. The history of travelogues throughout different periods of Baghdad's history is highlighted, with a particular focus on 18th and 19th century travelogues. This period was a critical epoch of change, not just in Baghdad, but across the world. Nevertheless, this book does not intend to provide a documentary of the travellers who visited Baghdad. It is rather an analytical study of the colonial literature in relation to the historiography of Baghdad.

Iman Al-Attar is an Iraqi architect, historian and a doctor in philosophy and urban history. She is currently doing research on topics including urban history, Islamic architecture, cultural issues and Baghdad. In 2018 she published her first book; *Baghdad: an urban history through the lens of literature*, which examines the writings of some local scholars who stayed in Baghdad in the 18th and 19th centuries.

Routledge Focus on Literature

Mapping the Origins of Figurative Language in Comparative Literature
Richard Trim

Metaphors of Mental Illness in Graphic Medicine
Sweetha Saji and Sathyaraj Venkatesan

Wanderers
Literature, Culture and the Open Road
David Brown Morris

Sham Ruins
A User's Guide
Brian Willems

The New Midlife Self-Writing
Emily O. Wittman

Female Physicians in American Literature
Abortion in 19th-Century Literature and Culture
Margaret Jay Jessee

Masculinities in Post-Millennial Popular Romance
Eirini Arvanitaki

A Glimpse at the Travelogues of Baghdad
Iman Al-Attar

For more information about this series, please visit: www.routledge.com/
Routledge-Focus-on-Literature/book-series/RFLT

A Glimpse at the Travelogues of Baghdad

Iman Al-Attar

Routledge
Taylor & Francis Group

NEW YORK AND LONDON

First published 2023
by Routledge
605 Third Avenue, New York, NY 10158

and by Routledge
4 Park Square, Milton Park, Abingdon, Oxon, OX14 4RN

Routledge is an imprint of the Taylor & Francis Group, an informa business

Library of Congress Cataloging-in-Publication Data
Names: Al-Attar, Iman, author.
Title: A glimpse at the travelogues of Baghdad/Iman Al-Attar.
Other titles: Routledge focus on literature.
Description: New York, NY: Routledge, Taylor & Francis Group, 2023. | Series: Routledge focus on literature; 1 | Includes bibliographical references and index. | Contents: The context of travel writing – Baghdad and the Ottoman Empire – Travelogues and history Representation – Diverse travelers before the 18th Century – Europeans Involvement in the 18th Century – British Intervention Intensifies – Summary. | Identifiers: LCCN 2022017045 (print) | LCCN 2022017046 (ebook) | ISBN 9781032188102 (hardback) | ISBN 9781032188119 (paperback) | ISBN 9781003326564 (ebook)
Subjects: LCSH: Travelers' writings, European–History and criticism. | Travelers–Iraq–Baghdad. | Baghdad (Iraq)–Description and travel. | Baghdad (Iraq)–History.
Classification: LCC DS79.9.B25 A45 2023 (print) | LCC DS79.9.B25 (ebook) | DDC 956.7/4703–dc23/eng/20220525
LC record available at https://lccn.loc.gov/2022017045
LC ebook record available at https://lccn.loc.gov/2022017046

ISBN: 9781032188102 (hbk)
ISBN: 9781032188119 (pbk)
ISBN: 9781003326564 (ebk)

DOI: 10.4324/b23141

Typeset in Times New Roman
by Deanta Global Publishing Services, Chennai, India

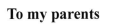

To my parents

Contents

Illustrations

Author's Note

All single dates are in Anno Domini (AD). Double dates are included to show the *hijrī* dates that are based on the Islamic calendar in the following format: AH (*hijrī*)/AD. In transliteration, I have followed the *IJMES* system. According to this system, diacritics are not added to personal names and place names. Diacritical scripts are used on technical terms, which are italicised and translated. Arabic terms known to English readers are italicised without translation.

Introduction

The study of the urban history of Baghdad in the 18th and 19th centuries has been a fascinating subject for me. Throughout my PhD study, I spent more than four years of my life researching different texts on Baghdad in the period between the mid-18th and mid-19th centuries. The expounded comparative analysis of these texts opened up a huge scope of understanding for me. Although it focussed on a specific period of Baghdad's history, it made me envision the whole history of Baghdad differently. Understanding history assists in recognising the present situation, in addition to introducing a future vision based on comprehensive acknowledgements and recommendations. Since urban history studies usually involve studies of different aspects of society, including social, political and commercial aspects, expanded knowledge of the nature of these aspects, in addition to recognising historical laws and the philosophy behind complex circumstances, improved my understanding of life in general.

In this book I am sharing part of this experience and hoping that the reader will enjoy it and gain similar impressions and reflections. I have decided to publish my doctorate research work in separate books, to help the reader to understand the concepts of each book thoroughly. These concepts emphasise collective analyses and integrated methods of reading history, favouring it over the usual sequential reading of events. The history of Baghdad had predominantly been written by two groups. The first group is Baghdadi scholars, and the second group is the travellers who visited Baghdad throughout different periods of its history. Collective reading of these sources provides a more accurate understanding of past events. It also delivers crucial information through the realisation of the critical role of the methods of interpretation, and how they affect the meaning of history greatly.

The conventional history of Baghdad mainly implements four methods of history writing. The first method is the 'personal or heroic approach', when history is totally associated with the history of prominent governors, military

DOI: 10.4324/b23141-1

leaders or a particularly effective group in society. The second method is the 'mythical and imaginative approach', which exaggerates historical events, and adjoins inventive ideas that drive them away from integrated methods of writing. The excessive astonishment of the round city of Baghdad and the exaggeration of the tragedy of the destruction of Baghdad are some examples of this technique. The third method is the 'physical or substantial approach' that involves documenting physical forms, building techniques and highlighting individual buildings' specifications. And the fourth technique of writing is the 'political scheme' that associates the history of Baghdad with personal views and political attitudes. The products of the colonial literature, or European travelogues, are the key example of this type since they enclose a particular focus on political issues. Nevertheless, because of the nature of writing and the interconnection of historical events, historical literature may combine more than one method, which makes these writings enriching.

The writings of some local scholars who stayed in Baghdad in the 18th and 19th centuries were discussed in my previous book; *Baghdad: an urban history through the lens of literature*. The significant literature produced by Baghdadis embraced plentiful themes of the urban development of the city, which have been overlooked in conventional historiography. These themes include the interlocking qualities of the Tigris River, historical settings of markets, multiple meanings of gardens, connotations of learning spaces and the social significance of houses.[1] Owing to relative stability, the 19th century specifically witnessed more active attempts in writing. Scholars became the most effective group; they helped to settle the society after each war or fatal epidemic. Those scholars were usually part of big families that were well known for their knowledge and high social status. Due to continuous changes in the political situation, some of these scholars moved to other cities in Iraq, with many of them composing brilliant writings about longing for Baghdad. Alternatively, other scholars chose either to stay in Baghdad for a short period, or to spend the greatest part of their lives there.

The constant movement of local scholars into and out of Baghdad influenced their writings. They enjoyed living around, and expressed a great deal of appreciation of Baghdad's environment, architecture and social life. They also conveyed sad emotions of sentimentality and longing, especially when they were away from it. Accordingly, the writings of Baghdadi scholars outlined significant nostalgic and spatial themes that were denoted in an expressive mode, indicating the spirit and character of the place and the people. As there was no requirement to specialise in a single discipline at that time,[2] those scholars were experts in multiple areas of knowledge, such as history, geography, astronomy, pharmacy, philosophy and theology, in addition to Arabic grammar and literature. Hence, their intellectual products contributed to the writing of the history of Baghdad in many ways.

Having investigated some available sources from inside in my previous book, this book explores some resources from outside, in order to reach a complete understanding of the history of Baghdad and its implications. The Europeans who visited Baghdad in the 18th and 19thcentury periods also provided significant writings. Yet their style was different from local scholars; it was empty of emotions and feelings, because of their different objective, which was mainly political. The role of travelogues is examined in relation to the representation of history in general, and the depiction of the urban history of Baghdad in particular. In addition, the history of travelogues throughout different periods of Baghdad's history is highlighted, with a specific focus on 18th and 19thcentury travelogues. This period is significant; it interlocks with the colonial era that has heavily influenced the modern phase of history. It also witnessed a lot of transformations, and it was a critical epoch of change, not just in Baghdad, but across the world. The urban development of Baghdad in this period was influenced by a number of social, geographical, environmental and political elements, such as the increased tendency to construct more roads, building and renovating mosques and schools, in addition to a complex administrative situation.

It is important to note that this book does not intend to provide a documentary of the travellers who visited Baghdad through different periods. It is rather an analytical study of the colonial literature in relation to the historiography of Baghdad. The investigation in this book aims at a critical reading of texts that takes into consideration all the circumstances of historical writing. In general, historiography designates the body of historical work on a particular subject, and the methods used by historians to present history, including specific sources, techniques and theoretical approaches. Since the issues of historiography are common worldwide, this book could assist in providing evidence of how history writing affects the perception of place in general.

The book starts with an analysis of the context of travel writing, followed by the accounts of the Ottoman Empire in relation to Baghdad. Chapter Three provides a reading of the role of travelogues in representing the urban history of Baghdad. A brief exploration of the history of travellers and travelogues of Baghdad before the 18th century is introduced in Chapter Four. In Chapter Five, the travelogues of two selected travellers in the late 18th century are examined, namely Niebuhr and Olivier. This is followed by a representation of the travelogues of three travellers in the 19th century, namely Rich, Buckingham and Heude. The work of other travellers could be discussed in further studies, as investigating every piece of colonial literature is beyond the capacity of this book.

Notes

1 These themes have been analysed in al-Attar, I 2018, *Baghdad: an urban history through the lens of literature*, Routledge, London and New York.
2 Compared with our present situation, where people at universities mainly focus on one major discipline.

1   The Context of Travel Writing

Generally, travel writing is influenced by a number of factors, such as the purpose of the visit, specific experience and circumstances of each individual, in addition to their knowledge and background. The travellers of Baghdad can be classified into two groups: travellers from the regional area of Baghdad, and travellers from Europe and other western countries. While the majority of European travellers arrived in Baghdad between the 17th and 19th centuries, regional travellers were interested in visiting Baghdad way earlier, since the time of the foundation of the round city in the eighth century. The number of regional travellers decreased from the 16th century onwards, due to floods, pandemics and an unsettled political situation. However, the 18th and 19th century periods also beheld a great number of scholars who lived in Baghdad and wrote their observations comprehensively.

Regional travellers who visited Baghdad at various times were mainly Muslim historians, teachers and writers, who would come from the regional area around Baghdad. At that time, especially in the earliest centuries that followed the foundation of the round city, Baghdad was the capital of the Islamic world, and it had many schools and learning centres that impressed many people, alongside its attractive natural environment and architecture. These characteristics, specifically being the centre of knowledge with eminent learning centres, made people, especially scholars, yearn to visit it either to study or to teach.[1]

Also, many people visited Baghdad on their way to Makkah, for example, or as part of their travel exploration in the Islamic world, to learn about different places and their history. This motive is a response to the Qur'anic teachings that advise people to travel to learn about other places and take lessons from different historical experiences. An example of these verses: *"Say (O Muhammad): travel in the land, and see the nature of the consequence for those who were before you! Most of them were idolaters"*.[2] A second verse advises people to travel and move to learn about Allah's

DOI: 10.4324/b23141-2

power and His capability of creating everything in this world: *"Say (O Muhammad): travel in the land and see how He originated creation, then Allah brings forth the later growth. Lo! Allah is able to do all things"*.[3] As exploration was the main purpose of regional travellers, this group can be considered 'explorers'.

It is interesting that the Arabic term *'siyaha'* which corresponds to the word 'tourism' and is normally associated with the act of 'travelling' these days, has only been mentioned once in the Qur'an. This verse orders the unbelievers who rejected Islam and did not comply to the terms of their treaties with Muslims, to move freely out of the Islamic lands, to avoid any consequences: *"travel freely in the land for four months, and know that you cannot escape from Allah's power and that Allah will confound the disbelievers (in His guidance)"*.[4] On the other hand, the Arabic term *'sayr'* that is the correspondent to 'walk or move' is mentioned more frequently in the Qur'an.[5] This implies the correct word for the action of travelling is movement' rather than 'tourism'.

On the other hand, the travellers who visited Baghdad between the late 17th century and late 19th century were mainly Europeans. In general, travellers in this period displayed a passion for travel everywhere in the world, as the primary means to learn about different places "while simultaneously perceiving travel narratives, history books, historical paintings, and architectural ruins to be modes of vicarious travel through time and space".[6] According to Chard, those travellers often "reflected on the pleasure that their commentaries might offer the reader".[7] Unlike regional travellers, European travellers were keen to entertain those who were interested in reading long narratives about remote places, and to provide information for a possible invasion of those lands. The task of providing long detailed information about the lands they visited made them less concerned if the information was fully accurate, or if collecting information and artefacts was ethical or not. Furthermore, because of their colonial ambitions, they approached these lands with superior, aggressive and racist minds. These attitudes planted the seeds for later European colonisation.

These issues are rarely outlined in contemporary discourse about European travelogues, which considers them documentations of mere travel experiences. The idea of colonising was strongly consistent with travelling. European travellers misguidedly considered racism against local people an indication of good national character, without expressing any signs of guilt. Simpson explains:

> The idea of traveling to the ends of the globe, visiting unknown and exotic realms, was almost beyond imagining, and to the reading public, therefore, there was an irresistible lure in accounts of such travels ...

[yet] few agonized over the rights or wrongs of colonizing other countries, most saw rivalry with other nations as a sign of vitality, and prejudice against the foreigner was seen as a healthy expression of national character.[8]

Since travel narratives were observations from outside, travellers relied on visibility and distinctiveness to conceptualise the city's image, its structure and context. Besides, political and psychological issues influenced their impressions greatly. During the 18th century, a number of European travellers passed through Baghdad, along with other adjacent cities like Damascus, Istanbul and Cairo. They produced drawings and recorded their observations about many aspects of the city. Yet their impressions were different from local scholars and regional explorers, because they viewed Baghdad with preoccupied minds and prejudiced images.

Consequently, their writings were clear representations of unclear visions of the city, as their passion for travel was often interrelated with specific political interests, and highly inquisitive attitudes. Also, spatial and temporal improvements in Baghdad enacted by their travel writings were outlined in relation to political issues rather than social or environmental issues. Accordingly, their narratives can be designated as personal travel diaries written by people who had a motive to record detailed observations about the places they visited and collect antiquities as a response to the desires of authority. These writings have been "more valuable for what they tell the reader about the personalities and ethnocentric leanings of their authors than for what they have to say about the people of the Middle East".[9]

Therefore, European travellers in this book can be referred to as 'emissaries'. Since they were abundant and provided detailed information about Baghdad, European travelogues had been among the important references for historical studies of Baghdad. Many of them were literally translated to Arabic. Nevertheless, the uncritical reading of these texts resulted in a noticeable miscount of some negative ideas presented in them. These ideas will be elaborated in the following chapters. The main components of the city were either misinterpreted or interpreted differently. For example, the Tigris River, which was normally the first component of Baghdad that travellers perceived as they approached it, attracted Baghdadi residents and the regional visitors to a great degree that made them compose poems to express their love and appreciation of the river's environment. Though the emissaries focused on the statistical information of the river, and hardly expressed their appreciation of it, which clearly shows the colonial objectives of their visits.

Among the shortcomings of the emissaries' texts is their material symbolisation of architecture as an artefact and art. This approach influenced conventional historiographies by stressing a limited set of architectural 'hero' figures.[10] In addition, aesthetic perceptions of these texts typically relate to the classical picture of cities in the Muslim world. Therefore, the city image drawn by them "fits naturally into the fundamental concept of Orientalism".[11] This concept has been strongly criticised by historians for providing a radical perception of the cities in the Islamic world. Simpson notes; "the long tradition of European hostility to Islam helped to foster a condescending way of thought that is now commonly described as Orientalism".[12] Before investigating these issues comprehensively, a brief examination of the history of Baghdad is essential, in the next chapter, to recognise the circumstances of Baghdad through different periods of history.

Notes

1 See al-Attar, *Baghdad*, pp. 25–26.
2 The Holy Qur'an, 30:42, <http://www.quranexplorer.com/quran/>, viewed 26 August 2021, Arabic script:
"قُلْ سِيرُوا۟ فِى ٱلْأَرْضِ فَٱنظُرُوا۟ كَيْفَ كَانَ عَٰقِبَةُ ٱلَّذِينَ مِن قَبْلُ ۚ كَانَ أَكْثَرُهُم مُّشْرِكِينَ" الروم 42
3 The Holy Qur'an, 29:20, <http://www.quranexplorer.com/quran/>, viewed 26 August 2021, Arabic script:
"قُلْ سِيرُوا۟ فِى ٱلْأَرْضِ فَٱنظُرُوا۟ كَيْفَ بَدَأَ ٱلْخَلْقَ ۚ ثُمَّ ٱللَّهُ يُنشِئُ ٱلنَّشْأَةَ ٱلْءَاخِرَةَ ۚ إِنَّ ٱللَّهَ عَلَىٰ كُلِّ شَىْءٍ قَدِيرٌ" العنكبوت 20
4 The Holy Qur'an, 9:2, <http://www.quranexplorer.com/quran/>, viewed 26 August 2021, Arabic script:
"فَسِيحُوا۟ فِى ٱلْأَرْضِ أَرْبَعَةَ أَشْهُرٍ وَٱعْلَمُوٓا۟ أَنَّكُمْ غَيْرُ مُعْجِزِى ٱللَّهِ ۙ وَأَنَّ ٱللَّهَ مُخْزِى ٱلْكَٰفِرِينَ" التوبة 2
5 The term '*sayr*' appeared four times in the Qur'an:
"*Say (unto the disbelievers): travel in the land, and see the nature of the consequence for the rejecters!*" 6:11
"قُلْ سِيرُوا۟ فِى ٱلْأَرْضِ ثُمَّ ٱنظُرُوا۟ كَيْفَ كَانَ عَٰقِبَةُ ٱلْمُكَذِّبِينَ" الأنعام 11
"*Say (unto them, O Muhammad): travel in the land and see the nature of the sequel for the guilty*" 27:69
"قُلْ سِيرُوا۟ فِى ٱلْأَرْضِ فَٱنظُرُوا۟ كَيْفَ كَانَ عَٰقِبَةُ ٱلْمُجْرِمِينَ" النمل 69
"*Say (O Muhammad): travel in the land and see how He originated creation, then Allah brings forth the later growth. Lo! Allah is able to do all things*" 29:20
"قُلْ سِيرُوا۟ فِى ٱلْأَرْضِ فَٱنظُرُوا۟ كَيْفَ بَدَأَ ٱلْخَلْقَ ۚ ثُمَّ ٱللَّهُ يُنشِئُ ٱلنَّشْأَةَ ٱلْءَاخِرَةَ ۚ إِنَّ ٱللَّهَ عَلَىٰ كُلِّ شَىْءٍ قَدِيرٌ" العنكبوت 20
6 Boyer, MC 1994, *The city of collective memory: its historical imagery and architectural entertainments*, MIT Press, Cambridge, p. 228.
7 Chard, C 1999, *Pleasure and guilt on the grand tour: travel writing and imaginative geography, 1600–1830*, Manchester University Press, Manchester, p. 2.
8 Simpson, M 1989, Orientalist travellers, Aramco World; 'Arab and Islamic culture and connections', vol. 40, July/August, no. 4, pp. 16–18. <https://archive .aramcoworld.com/issue/198904/Orientalist.travelers.htm> viewed 4 March 2020.
9 Simpson, 'Arab and Islamic culture and connections'.

10 Nasar, J 1998, *The evaluative image of the city*, Sage Publications, Thousand Oaks, CA, p. 21.

11 Raymond, A 2002, *Arab cities in the Ottoman period: Cairo, Syria, and the Maghreb*, Ashgate, Variorum, Aldershot, Hampshire, Great Britain; Burlington, VT, p. 3.

12 Simpson, 'Arab and Islamic culture and connections'.

2 Baghdad and the Ottoman Empire

In order to understand the circumstances of Baghdad during the 18th and 19th centuries, it is important to recognise the history of Baghdad in general. It is equally important to study the history of the wider region of Baghdad, and the different social networks with which it was connected, to understand the background of the historical events and uncover some accounts that are misinterpreted or underrepresented in conventional histories. Hence, a brief study of the history of the Ottoman Empire is also essential to this inquiry.

One Name and Multiple Locations

The history of the dwelling in the Baghdad area goes back to 4,000 years ago.[1] It had been a village to the east of the Tigris River, and a marketplace for a long time. Yet in the year 144/762 a nearby site, on the western side of the Tigris River, was chosen by the Abbasids to build a round city to be the capital of their state.[2] The construction of the round city of Baghdad started that year, and it concluded five years later, in 149/767. This city was the first planned city in the history of Baghdad. The round city's status and its fascinating building techniques attracted many people and caused a noticeable increase in the number of immigrants to Baghdad. As a result, Baghdad became a famous international city. However, shortly after its completion, the suburbs of the round city started to extend outside, due to the rigid round shape and strict political rules.

In 221/836 the Abbasids abandoned the round city; moving their capital to Samarra' north of Baghdad.[3] The round city eventually started to decay until it vanished completely with time.[4] Extending the buildings outside the round city connected it to another urban town that was also developed during the early Abbasid period.[5] This town was called *al-karkh*, situated on the western bank of the Tigris, and it reached out to the original village on the eastern bank of the Tigris that was later called *al-rusafa*.[6] In 278/892 the Abbasids moved their capital from Samarra' back to the Baghdad area.[7]

DOI: 10.4324/b23141-3

However, they chose the eastern side of the river near the ancient market-place. The second Abbasid capital was also called Baghdad and it survived until today and it was also called Baghdad. The whole area to the east of the Tigris River was later called *al-rusafa* (Figure 2.1).

The second Abbasids' settlement of Baghdad comprised the centre of the city's important divisions, markets and businesses. This locale continued to be the central hub for successive governments after the Abbasids, including the Ottomans, and it has been the commercial core of Baghdad for centuries until the present day. In 334/946 Persian Buyids controlled the western side of Baghdad, in addition to a few areas on the eastern side of the city. In 443/1052 Turkish Seljuks took over parts of Baghdad.[8] The Abbasids took back Baghdad in 582/1187, but in 683/1258 the Mongols invaded it and put an end to the Abbasids' rule. After the Mongol invasion,

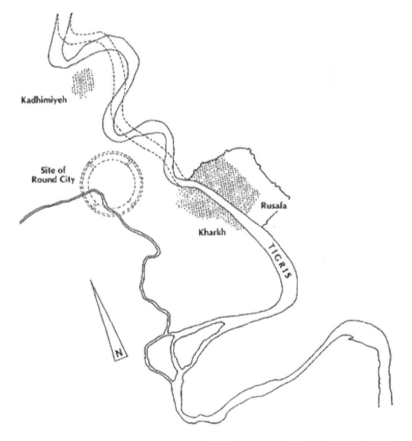

Figure 2.1 Round city site and the second Abbasid capital [Warren & Fethi 1982].

Baghdad was controlled by a number of empires. In 802/1400 the Jalayirid controlled Baghdad, and in 813/1411 the Turkmen occupied it again. In less than 100 years the Safavids controlled it in 913/1508, and in 940/1534 the Ottomans took over Baghdad. The Safavids took it over again in 1030/1621 until 1047/1638 when the Ottomans regained their control over it.

In 1116/1704 a Mamluk *pāshā* called Hassan *al-Jadid* started to govern Baghdad. At that stage, he was more connected to Ottoman *sultans*. Yet the Mamluks' official rule of Baghdad started in 1136/1750, and it lasted in 1246/1831 when the Ottomans controlled Baghdad completely. In 1335/1917 the British occupied Baghdad, until 1339/1921.[9] This brief illustration of the political history of Baghdad shows the transitoriness of different historical accounts, and the temporality of historical events, which is one of the main characteristics of history.

A Unique City with a Mysterious History

The history of Baghdad in the 18th and 19th centuries holds some enigmatic aspects, due to the shortage of historical materials from this period, and the misinterpretation of the events that occurred then, down to political reasons. Selen Morkoc suggests the texts and architectural narratives inherited from the Ottomans are challenging because they are "from a world that is remote both in time and culture".[10] The slight unfamiliarity with those texts belongs to the big gaps generated by modern attitudes and the emotional effects of modernity, which shaded the Ottoman time with oldness and antiquity. Besides, this period is not that far away in reality, which makes this suggestion invalid.

Another reason for the enigmatic history is suggested by Andre Raymond. He notes; an investigation into the historical sociology of Baghdad shows that the reason behind the mysterious situation and the lack of documentation of Baghdad's urban development in the 18th and 19th centuries is the pre-eminence of French scholarship at that time, who did not pay great attention to 'British-controlled Iraq'[11] European travellers who visited Baghdad in the 18th century were mostly German and French, while the travellers of the 19th century were mostly British and American. The gradual change of French interest in Baghdad in the 18th century may have caused a lack of literature by international writers. Still, the texts of local scholars who documented some events could compensate for this shortage.

On the other hand, Michael Bonine suggests that despite the continuous damage to the original documents, the available documents provide adequate information about the city's history. He implies that the lack of awareness of the 18th and 19th century texts of Baghdad is a result of political tensions which may have curtailed the research and made it difficult to progress further.[12] The case of Baghdad is complex if it was envisioned from political perspectives, yet undoubtedly complex situations affected the literature movement to some extent.

Another reason for the enigmatic history of Baghdad is that the European travellers approached it with an imaginary picture of the round Abbasid city that was built in 144/762. When they reached the 18th century city they encountered a different picture, which caused confusion that was reflected on their writings. The round city of Baghdad was carefully planned and heavily decorated in order to provide a high level of security and protection to the caliph. These luxurious qualities attracted the Orientalists a lot. Although no single trace is left of this city, the memory of its original shape and the ideas behind it lasted for centuries in a way that has no parallel in history. This unrealistic vision occupied the minds of the European travellers and placed it in constant comparison with the situation of the city at the time of their visit.

In addition to the aesthetic criteria and power expressed by the round city theme, the perfect geometrical shape that resembles some ancient cities in Mesopotamia like Seleucia contributed to the admiration factor. Grabar declares; "the perfect astonishing urban composition of Baghdad is not really an urban one but a palatial one to which none of the early Islamic cities corresponds".[13] Some historians consider it a utopian experiment in a scientific order,[14] placing it as a model of Islamic cities. This approach increased worldly extensive studies to discover more about it as a beautiful medieval city that signifies 'works of art'[15] In fact, the historical narratives that were written according to the requests of the Abbasid caliphs exaggerated the outstanding criteria of the round city. Nevertheless, its complete destruction contributed to the uncertainty of its real image. It is interesting how the round city myth served as an example of a psychological tool that influenced arts, culture and historiography to a great extent.

A further reason for Baghdad's enigmatic history is having had multiple locations with the same name; Baghdad. Examples of these locations are the old marketplace village, the Abbasids' round city, the administrative hub in eastern Baghdad and present-day Baghdad that comprises all these locations, in addition to *al-karkh*. While some of them were the centre of attention at some time in history, others had similar status in different times. For example, the compassion of the local poets was on Ottoman Baghdad, yet conversely European travellers admired the round city a lot and underestimated the city of the 18th century because of some negative visual appeal. Nonetheless, the solid image of the historical city became abstract with time when it experienced meaning-loss because of the lack of profound interpretation due to the advancement of modernity. This situation could be one of the reasons for the mysterious history; it makes the case of Baghdad as an exemplar of how superficial visions by outsiders critically affected its urban development and growth.

These issues might have also affected the progress of anthropological and sociological studies. The extensive conservation efforts in the last decades maintained individual historical buildings physically, but they were not successful in maintaining the whole historical city and its meaning. However, what makes Baghdad distinctive among other cities in the region is its unique position that allowed a significant mixture of people from diverse backgrounds. Also, its perseverance and ability to survive serious occurrences throughout history established this sense of distinctiveness. In addition, the historical position of Baghdad as a marketplace and hub of knowledge and learning for centuries increased its uniqueness and established a great attachment within people that has been reflected in literature and poems.

The ancient history of Baghdad enabled it to inherit a unique urban heritage from different phases in history. For example, it inherited a unique housing style that utilised exceptional techniques to fulfil social requirements and introduced brilliant solutions for inconsistent weather conditions (Figure 2.2). These methods made it one of the most successful aspects of Baghdad's urban history that has continued to inspire architects and builders until today. Historical documentation of the houses that survived through the 19th century point out that the unique style of those houses was inherited from the 18th century or prior to that time,[16] yet the resources focus on the great appeal of the 19th century houses, because of their extensive ornamentations and decorations (Figure 2.3).

The Ottoman Empire

The subject of Ottoman history has increased notably in current historical studies. These studies are part of the social research promoted in conventional historiography, to provide specific links to the critical moments of transformation in the cities that were under Ottoman control. Such studies offer insights into the historical, geographical, economic and political circumstances of the period preceding the 20th century. Though historical references for the studies of the Ottoman period are many, the history of this period remains unclear. It appears that the cause of the vagueness of this history has more to do with methodology than with limited data. For instance, most historians have a tendency to situate the subject of their study within administrative and political accounts rather than economic and social histories.

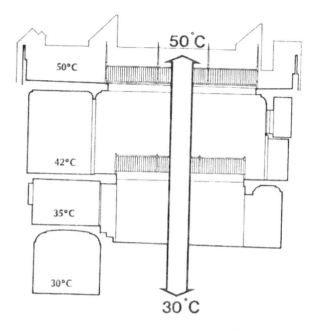

Figure 2.2 Extensive weather techniques in Baghdadi houses [Warren & Fethi 1982].

Figure 2.3 Beautiful ornamentations in a Baghdadi house [Warren & Fethi 1982].

The prevailing attitude of the writings that are concerned with the architecture of the Ottoman period reflects a sense of antiquity that generates some pride for Turkish historians. The arguments proposed by these studies provoke a feeling that the ideas that are correlated to that period are usually located in a world that does not exist anymore. However, this distance in time is not factual. The focus of these studies on specific terms that are not the centre of present studies, such as Turkish towns, Arabic towns, *harem*, *ulema* and *waqf* asserts the feeling of remoteness and unfamiliarity. This tendency widens the gap between historical events and creates barriers and challenges to the understanding of history. It is inevitable that history is a connected chain of events that cannot be separated. Thus, modern history should be studied in connection to the preceding accounts, to promote more understanding of history, and to maintain the advancement of present and future times.

The Ottoman Empire was one of the largest and longest-lasting empires in history. It was founded in 1299 and ended officially in 1922. The Ottomans were originally an ethnic group living in East Asia. Later they moved to north-east Anatolia where they founded the empire under the leadership of a strong warrior called Osman, from which the name Ottoman is derived.[17] The highest authority in the Ottoman Empire is usually the *sultan*, who was strictly obeyed: "God is watching, as also is the Sultan".[18] The goal of establishing the Ottoman Empire was quite parallel to other empires in history, which was basically political. The Ottomans were officially Muslims, though some laws that were implemented in the empire were not fully derived from Islamic principles, either because of ignorance, or because of specific personal or political leanings.

On some occasions, the decisions of the *sultan* or other officials were contradictory to Islamic laws, which suggests the effect of other incentives on these decisions.[19] The irresponsible administration of taxes, unjust orders and appointing people from other religions as high officials, are some examples of those policies. Regardless, conventional historiography associates the Ottoman Empire with Islam, which is superficially true, but deeply problematic.

The authority of the Ottomans in the Baghdad area was intermittent. They ruled the area between the years 1534 and 1621, and then controlled it between 1638 and 1704. The third period of their direct rule of Baghdad was between 1831 and 1917.[20] These consecutive interruptions of the same ruling power of Baghdad resulted in unstable conditions, which created political strain and weakened its urban development. The impact of irregular administrations is recognised in some of the literature of this period.

The Perception of Ottoman Cities in Historiography

The history of cities that were under the Ottoman Empire rule is important to this study, since Baghdad was among them. Historians consider the history of the 18th century a complex history and an "abstract and brief chronicle of the past"[21] The huge change that occurred around the world due to modernisation and industrialisation established this ambiguous vision of this century, because it lies on the edge of modern history. Current studies of Ottoman cities in the 18th century are centred on questions of aesthetics and affluence, such as arts and entertainment. These studies also focus on the 18th-century issues of public life and concern for public order that "never ceased to overlap".[22] In addition, historical sources always interrelate social issues with government settings rather than public affairs. Thus, they can be considered contributions to institutional history rather than social and urban history.[23] However, a significant shift in attitudes towards general urban affairs is evident in some sources. The study of Ottoman towns by Veinstein is an example of these approaches.[24] He questions the survival of a unique type of town characteristic of the Ottoman Empire that could naturally be labelled as 'Ottoman Town', and he argues that Istanbul *sultans* developed a proactive urban policy that was implemented in various ways and at different levels.[25]

Ottoman cities experienced specific conditions due to their integration into the empire. However, Veinstein suggests there was a policy of differentiation between different cities controlled by the Ottomans. Those conditions benefited specific towns unevenly, depending on the position of each town in the political, administrative, strategic and economic systems of the empire. He states Istanbul was privileged, being both the main centre of attraction for the empire's commercial exchange and the primary object of stately concern in terms of settlement, supplies, facilities, developments and beautification. Accordingly, he proposes the greater or lesser distance from the capital as a main differentiating factor between these towns.

Further, Veinstein suggests another reason for this disparity; a 'dividing line' that places Arabic provinces on one side, and Anatolia and non-Arabic areas on the other, which explains why historic Middle Eastern cities rank far behind Istanbul in the classification of the empire's main cities.[26] Veinstein also refers to the impact of various attitudes in historiography as another differentiating factor. He states that despite having been considered primarily different, Arabic and central provinces "have been studied rather differently by different people".[27] Such preconceptions impacted the understanding of these cities and contributed to their different perceptions in history. It is acceptable that distance might have been a distinguishing measure due to difficult transportation circumstances at that time. It is also reasonable that Ottoman *sultans* gave considerable attention to the architectural development of the capital of the empire, Istanbul, more than other cities. However, despite being Arabic, cities like Aleppo and Damascus were among the Ottoman territories that were "strikingly urbanised, when compared with Europe",[28] which eliminates a direct correlation between distance and significance.

These cities could have maintained a great level of prosperity, yet, according to some historians, they lost relative importance during the 18th century with the drying up of the Iranian silk trade.[29] This suggestion challenges the propositions of distance and language as the main factors of differentiation. Therefore, there have been other factors in the degeneration of these cities apart from the proposed points. The same thing can be assumed about Baghdad, which experienced some damage in the 18th century, and thus became less appealing for the Europeans compared with other cities. In addition to the factors mentioned above, the change in Baghdad's structures was due to floods and constant fights, which overlooks geographical proximity and ethnic diversity from the equation. During the 18th century, the Baghdad area was larger than the present city; it consisted of a number of cities that currently have separate municipalities. At that time, Iraqi land was incorporated into three main provinces: Baghdad, Basra and Mosul[30] (Figure 2.4). Abdullah refers to the impacts of political policies on the insight of these cities; "the empire's provinces of Mosul, Baghdad, and Basra were often regarded as significant only in a military sense".[31]

Despite describing them as fundamentally different, Veinstein suggests the integration of different cities into the immense Ottoman structure was a sign of prosperity, and this integration brought relative order, security and unified legislation.[32] In terms of art and architecture, those measures contributed to a certain standardisation of production, though different situations provoked obvious divergences in architectural styles and terminology. Consistent similarities with relative diversity in the art and architecture of cities controlled by the Ottomans are evident in the heritage of those cities. This aspect challenges the earlier suggestion of differentiating measures between those cities. Once again, Veinstein generalises Ottoman cities' affairs, contrary to the earlier proposal of differentiation. This case indicates the capacity of historiography for mixed suppositions in analysing historical accounts. Undoubtedly the cities under Ottoman control occasionally shared similar social and architectural features. The similarities were the consequences of the parallel laws enforced on those areas and easy movement between different regions in

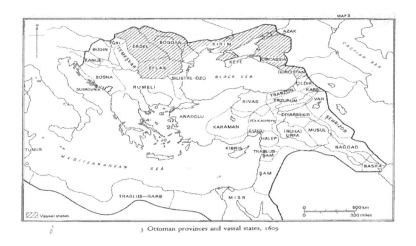

3 Ottoman provinces and vassal states, 1609

Figure 2.4 The Ottoman Empire and the provinces of Iraq [Inalcik & Quataret 1997].

the empire. Yet the divergence was mainly a result of diverse political circumstances.

While historians often relate the signs of order in Ottoman cities to the Islamic rules, Veinstein inaccurately states that unified measures were "stimulated by developing relations with the West".[33] He refers to the municipal reforms in the Ottoman Empire that came in the context of reform laws that began in the 1830s. The examination of the events at that time, and the reading of various historical sources, assert that many achievements were inspired by enthusiastic local people, and the terms 'reform' (*islahat*) and 'reform laws' (*tanzimat*), were among the tricks of the Europeans, who utilised these terms to interfere in the administration system, to arrange for a future control of the area. In addition, the word 'modernisation' was introduced by the Europeans to change the system to suit their plans and enhance their superiority. At this stage, European emissaries played a significant role, to ensure the implementation of these plans.

In her discussion about the *tanzimat* urban historiography, Nora Lafi asserts that those reforms were partially influenced by the Europeans, but she argues they were not importations onto a blank canvas, because "there was in every Ottoman city an old regime urban system".[34] The reforms were characterised by various attempts to 'modernise' the Ottoman Empire. Lafi indicates Europe in the mid-19th century was not an island of modernity, and French historiography "has also insisted on the importation tool".[35] The unique urban system in the Ottoman Empire is associated with Ottoman cities before the 19th-century reforms. This system is a result of thousands of years of urban development in the area. Thus, it was certainly a point of pride for the inhabitants, and for what Lafi calls the 'notability' who carried the urban government in their hands.[36]

Nevertheless, Veinstein points out some similarities between Ottoman towns, noting "all Ottoman towns undoubtedly share common features, such as less continuous circulation zones geometrically and topologically and more regular plots which emerged over time as private housing and concerns".[37] Finally, he expresses admiration of the urban change towards regular plots and constant zones. These statements indicate a lack of appreciation of traditional urbanism of those cities, with a promotion of the globalising motivation of urbanism schemes of the 20th century. This subject is among the extremely destructive issues of colonialism. It changed the urban system drastically, replacing the unique urban system that was adopted by the residents who developed it over thousands of years, with an imposed system that did not fit with the history of the place, its environmental and social requirements.

The Mamluks' Rule of Baghdad

The history of Baghdad during the Mamluks' rule is important for this book, since all the selected travellers visited Baghdad in the same period. Historians consider the second half of the 18th century the starting point of the autonomous Mamluks' rule of Baghdad, with the beginning of the rule of Sulayman abu Layla (1749–1762).[38] Though, in his book *tārīkh Baghdad*,[39] Sulayman Fa'iq implies that the groundwork of the Mamluks' rule of Baghdad started in the early 18th century with the appointment of Hassan *pāshā al-Jadid* between the years (1116/1704–1136/1723).[40] This *wālī* launched a development scheme by repairing shrines, caring for public gardens and amenities of the city and trying to attract back to Baghdad the people who deserted it for tribal refuge because of constant fights and disasters.[41]

Al-Suwaidi portrays an enhanced image of Baghdad during Hassan *pāshā*'s rule. He states the collapsing walls of Baghdad were renovated along with the ditch, to improve the city's defence lines. Also, some mosques were renovated, and new khans were built on both sides of Baghdad. He also points out that the construction movement hastened after 1716 because many people from neighbouring Iran took refuge in Baghdad and other cities in Iraq due to the extreme situation in their motherland, which was caused by famine and sieges. He notes the new immigrants to Baghdad found a better place in a city that was full of orchards and was desired by many people. And he designates some economic growth in this period that resulted in amending two heavy taxes in 1717.[42] However, in the palace, corruption started to appear as signs of prosperity and luxury.[43] Towards the end of Hassan *pāshā*'s rule, plague occurred in Baghdad in 1720, and many people were killed, either caused by the epidemic or famine. This period witnessed the establishment of a special office that brought the Mamluks to Baghdad. This office was responsible for training officers and hiring them to control the situation in Baghdad.[44] The matter of bringing more Mamluks to Baghdad and appointing them to important positions reflects the mistrust in local people. It also marks the starting influence of the British, since the Mamluks were allied with them.

Another big issue associated with the Mamluks' rule, which was intensified during Hassan *pāshā*'s rule, was the constant fights with the tribes that refused to pay taxes. European travelogues emphasised this issue; they harshly accused the tribes of being disobedient, labelling them as rebellious. The issue of taxes is a complex issue in the history of Baghdad. People may refuse to pay taxes because they are not satisfied with the government and its legitimacy. In other situations, the taxes may have been imposed on people in an unjust way. So, they may have had the right to refuse to pay taxes. Therefore, it is crucial for historians to examine this issue comprehensively before making any unfair judgement. The fights over taxes affected the city structure and the natural environment severely, causing great damage, owing to the careless attitudes like cutting trees or throwing robes or stones into the rivers. These matters have never been raised in conventional historical accounts, which reads the situation superficially.

The Mamluk *wāli* Ahmed *pāshā* succeeded his father Hassan *pāshā* in 1136/1723. His rule of Baghdad was an extension to the policies of his father. This period witnessed many disasters, like epidemics and floods, as well as extensive fighting with the Safavids and their successors, which resulted in a number of sieges and famine in Baghdad. Ahmed *pāshā* continued to build a powerful military-administrative apparatus based largely on importing soldiers from Georgia, who were later known as Mamluks.[45] The tendency to build and renovate mosques and schools was maintained in this period, which increased the number of visitors from the neighbouring cities and from Europe. Among those travellers are shaykh Mustafa bin Kamal al-Din al-Siddiqi from Damascus in 1726, shaykh Muhammed bin Aqila al-Makki in 1732 and the German traveller Carsten Niebuhr in 1733.[46]

Though Hassan *pāshā* and Ahmed *pāshā* were both appointed by the sultan, they were planning a gradual disconnection from the Ottomans' authority,[47] so they initiated the early steps towards political autonomy from the Ottomans. Those governors were bound by power and personal desires, and they had good relations with the European authorities, which brought political problems to Baghdad and the region. During the rule of Ahmed *pāshā*, which ended in 1162/1749, the first European consulate was established in Baghdad, namely the French Consulate. The strong relationships and alliance between these governors and the Europeans explain the huge number of positive comments by European historians about them. For instance, Longrigg associates strategic 'changes' in Baghdad's history with these governors. He calls the wars led by Hassan *pāshā* 'the wars of giants', and he praises this *wālī* and his father for overcoming external problems.[48] He uses the word 'discipline' unfairly when describing the fights with the tribes for tax collection, and he relates to the social characteristics of Baghdad incorrectly, describing it as a 'tribal society'[49], which gives the impression of backwardness. Whilst a number of tribes have lived in Baghdad and preserved their tribal traditions, they lived homogenously, and their social and urban traditions were transformed gradually to achieve social harmony.[50]

The word 'discipline' and its corresponding term in Arabic, *ta'dib*,[51] have been used frequently in other historical texts by European authors. The heavy use of this word indicates that these linguistic techniques were utilised to serve political plans. The dangerous aspect of such comments is that they are written by a foreigner who understands little about Baghdad, and whose observations were influenced by political concerns. Since they were not grounded on a comprehensive analysis of the society, these comments did not reflect the real situation of Baghdad. This issue encapsulates one of the flaws of conventional historiography, which often mixes real events with imagined or invented matters, causing confusion by making it challenging to understand the real historical situation.

On the other hand, Fa'iq states that both Hassan *pāshā* and Ahmed *pāshā* have communicated with the Ottoman *sultan* more than Sulayman 'abu Layla, who succeeded them. He notes Sulayman 'abu Layla, the son-in-law of Ahmed *pāshā*, was determined to improve the situation in Baghdad, and sustain the same policies of the preceding rulers by intensifying tribal fights to increase peace.[52] It is contradictory and unreasonable to associate fights with peace, to please the rulers whom he describes as 'grand governors', especially if the fights are hostile. As discussed earlier, unfortunately tribal fights over tax collection are a multifaceted issue, and dealing with them incorrectly contributed to a gradual weakening of the city. These statements by Fa'iq show some similarity with Longrigg's comments, which indicates there was a political agenda behind both writings. While Longrigg had a colonial mentality, Fa'iq had strong connections with the governors, so he wanted to please them by praising their achievements, with no sympathy to local people. It seems that the constant focus on individuals and events, and discussing them from the governors' viewpoints only, has been the main style of the historiography of that period.

Another issue that continued to be the focus of this period's historiography is bringing a large number of Mamluks to Baghdad and training them to work in the government's administration office. Similar to Hassan *pasha*'s policies "under ʾabu Layla the use of Georgian freedmen in important posts was much increased".[53] Faʾiq suggests that Sulayman ʾabu Layla encouraged this policy because of his strong inclination towards those of his ethnicity.[54] Bringing more Mamluks to work in Baghdad undoubtedly changed the demographic map of Baghdad. In addition, it created two different social classes; Mamluk allies and city dwellers,[55] which widened the gap between the society and the government.

After the death of Sulayman ʾabu Layla, his deputy[56] Ali *pāshā* took over his position. Ali *pāshā* wasn't successful in this position; he was extremely mistrusted and was killed after two years by his deputy Omar *pāshā* (1764– 1775) who succeeded him.[57] According to Faʾiq, the situation worsened in this period with an increase in tribal uprising and with the plague outbreak in 1186/1772, which resulted in a great loss of scholars and highly educated people.[58] Subsequently, a huge amount of expressive literature and poetry was composed by the scholars who survived the plague. Those prose works contained important insights into Baghdad's urban environment and history.

The historian and religious scholar ʿabd al-Rahman al-Suwaidi wrote a book to document the events of the plague that occurred between 1772 and 1778. Although al-Suwaidi praised Omar *pāshā* with a poem when he was appointed as a governor of Baghdad, he was disappointed with this *wālī* after the plague.[59] Similar to Ali *pāshā*, Omar *pāshā* was killed in 1775 by another governor called Mustafa *pāshā*, who was appointed by the *sultan*. The new governor was selfish and had little knowledge about the distinctive criteria of Baghdad and Iraq in general. After one year of his rule, he was replaced by ʿabd Allah *pāshā* (1776–1777) who was also passionate about amusement and entertainment more than anything else.[60]

Historians who wrote about this period narrate that the treasury of Baghdad was extremely full, and the officers who were sent to Baghdad during that time were astonished with the high level of prosperity and wealth. The story of Selim Sirri, who was sent to Baghdad to solve a big problem is narrated by a number of historians, citing that as soon as Selim entered Baghdad, he immersed himself in fun and entertainment and forgot the task that brought him to Baghdad.[61] Those foreign Mamluk governors and officers collected massive amounts of money and became very rich, unlike the original people of Baghdad. Ideally, a governor of such a great city as Baghdad should be wise, knowledgeable and sincere, so he served the people wisely. Yet these governors represented the opposite. While at face value they constructed and maintained many buildings, they destroyed the society by wasting its wealth and its valuable heritage.

The pattern of short-term governors continued, which points out their great corruption. In 1191/1777 'abd Allah *pāshā* died, and consequently a civil war started in Baghdad down to the clashes over the governor's position. This war lasted for five months, and one can imagine how much fear, instability and destruction were caused by those selfish people. Eventually, another governor called Hassan *pāshā* (1778–1779) was appointed by the *sultan*. Again, this *wālī*'s ruling of Baghdad was short, and he was busy with entertainment and with fighting the tribes for taxes. According to Nawras, the residents of Baghdad were so upset with those continuous fights, so they erected dividing walls for security, and they attacked the government quarter, or *al-sarāy*. Hassan *pāshā* was able to escape from the government quarters to the other side of Baghdad (*al-karkh*) but he caught a disease and died shortly after.[62]

The short-term governor's age came to an end with the appointment of Sulayman *al-kabir* (the old), who ruled Baghdad for a long period; about 22 years (1780–1802). This governor gained a great deal of attention in historiography because he was supported by the British. Before his appointment as a governor of Baghdad, he was the governor in of Basra in 1779.[63] After some troubles and a siege on Basra by the Iranians he was arrested and imprisoned in Iran for a few months. Fa'iq states that the British Consulate in Shiraz interceded successfully for him with both the Ottoman and Iranian governments until he was released.[64] Then the British representative in Basra supported Sulayman and sent money to Istanbul to get approvals from the *sultan* and his head officers to appoint him as a governor of Baghdad.

Nawras notes the nomination of Sulayman *al-kabir* by British officials was an expansion of their long relations with him since 1765.[65] History writing that is mainly based on European travelogues of this period, demonstrates an enhanced atmosphere in Baghdad with the appointment of Sulayman *al-kabir* as a governor of Baghdad. This is yet again one of the tricks of historiography, which presents an improved image of him because he had good relations with the British Consulate. Nawras describes the ruling period of this *wālī* as a period full of 'massive events', encompassing corruption of the administrative system, intense tribal upheavals, and dangerous Wahhabi attacks.[66]

According to Fa'iq, Sulayman *al-kabir* tried to restore order and defeat corruption on both internal and external levels. An example of this corruption: he narrates a fight over the deputy governor's position, which created a lot of tension and disarray in the community. He states the treasury of Baghdad was full of money because expensive goods were collected from the merchants and farmers. A large amount of money was sent by the *wālī* Sulayman to Istanbul each year to please the *sultan*. Faʾiq narrates when Sulayman *al-kabir* died in 1802, his deputy Ali *pāshā* was astonished at the great amount of money that had accumulated in Baghdad's treasury. As a deputy governor, he wrongly thought he owned this wealth, so he distributed part of it to the community to gain their recognition.[67]

These accounts show the unjust policies of Sulayman *al-kabir*, and the unstable conditions due to excessive political troubles that continued in this period. Despite this, Longrigg expresses a satisfaction with the ruling strategies of this *wālī*. He notes: with his entry to Baghdad he "opens the golden age of the slave-government of Iraq".[68] Longrigg suggests the long misrule of the Mamluks "sprang from the absence of will to govern well".[69] He adds, "in permitting the British Resident to become the second man in Iraq the Mamluk *pāshās* had shown some recognition of the means of progress, some willingness to be guided, some lightening of prejudice, occasional friendship and courtesy"[70]. So, he connects 'governing well' to giving permissions for a permanent British agent to be appointed in Baghdad in 1798, considering it, in his view, a great achievement.

Longrigg notes thereafter, "Baghdad became the chief centre of British influence".[71] Commending the governors who supported British benefits clearly reflects the strong will of the British to intrude and occupy the area. Despite all these statements that reflect invading interests, Longrigg's book was translated into Arabic and became among the sources of the history of that period, which makes it necessary to question such historical references before implementing them. The long rule of Sulayman *al-kabir* ended in the early 19th century, and the governance of Baghdad transitioned to the hands of short-term governors who could not achieve a lot during their reign.

When Sulayman died in 1802, Ali *pāshā* was appointed as a governor of Baghdad. He continued the same policies of Sulayman, and he also fought the Wahhabis, and he was killed in 1807. Sulayman *al-saghir* (the young) succeeded him in 1807, but he was killed in 1810, and then followed by ʿabd Allah *pāshā*, who was killed in 1812. In 1813 Saʿid *pāshā*, son of Sulayman *al-kabir*, became the governor.[72] As noted by Izz al-Din, Saʿid, who was only 22 years old, was a weak governor and was guided by his mother. He spent large amounts of the treasury of Baghdad on his entertainment.[73] In 1816 Saʿid was killed, and Dawud *pāshā*, the son-in-law of Sulayman *al-kabir*, was appointed as a *wālī*. He remained a governor of Baghdad until 1831[74] when the Ottomans sent a military force to Baghdad to put an end to the Mamluks' rule in Baghdad.

Apart from fluctuation and instability, the early decades of the 19th century witnessed few remarkable changes in both social and urban development of Baghdad (Figure 2.5). Al-Warid narrates that a market complex was opened in *bāb al-agha* in 1802.[75] Also, in 1821 the first stone press was established in *al-Kazimiyya*, north Baghdad, and in 1830 the first mechanical press was established in Baghdad, namely '*maṭbaʿat dār al-salām*'.[76] In addition, new roads were opened in 1825 on both sides of the bridge.[77] On a broader level, 19th-century Baghdad experienced more urban sprawl following the gradual destruction of the city walls and the constant fights inside and outside *al-sarāy*, which made escaping the troubles of the central administrative area a favourable option. People started to move from eastern Baghdad outside the wall, across the river to the western side (*al-karkh*). This movement contributed to the urban growth of the western part of the city.

Longrigg associates these changes with the ruling of Dawud *pāshā*[78], stating that Dawud's support for learning, his generosity, and his frank 'independence of Istanbul' made him more popular.[79] Similarly, Dina Khoury notes there are three important social and political developments in Baghdad in the 19th century: the rise of Dawud *pāshā* as a governor of Baghdad, the increasing power of religious scholars and urban elites in the economic and social life of Baghdad and the gradual subjugation of other localities and political centres in Iraq to Baghdad.[80] Yet again, historians exaggerate the achievements of a governor and relate the urban growth to his rule, which is not correct, because the rule of Dawud *pāshā* was not much different from that of other Mamluks. For instance, he instigated violence and fights to collect taxes, which affected the society badly. Besides, during his rule, estates were freely bestowed without restrictions.[81]

Figure 2.5 Nineteenth-century Baghdad [Makkiyya 1968].

Because the 19th century had links to the reform plans, its psychological impression as "one of the exceptional changes in Ottoman social, economic and political life",[82] drew an improvised picture of that period in historiography. Undoubtedly 19th-century Baghdad experienced some urban growth. Though, historiography inflated the achievements of the governors, and extended a monolayer interpretation of history that focuses on a single factor only. If we view Baghdad in the 19th century from architectural and urban development aspects, with no consideration to other matters, it could appear as a time of remarkable developments, whereas political plans were the unseen motives for many reform plans. Unfortunately, the main concerns of historiography are politically related issues rather than other matters, which asserts how specific attitudes in historiography could greatly impact the understanding of history.

While historiography positively focuses on the achievements of the Mamluk governors and on the principles of reform as signs of stable and peaceful conditions, Baghdad's situation was always tied to trouble. Nawras explains there was not one year which passed without turmoil, fear and famine. He adds, out of the total 11 Mamluks who ruled Baghdad, six of them ended being killed, which reflects the massive chaos and internal conflicts during their rule of Baghdad.[83] The adverse attitudes of the Mamluks placed Iraq "in an intermediate position between Syria and Egypt, sharing with Syria the problem of the tribes, sharing with Egypt the Mamluks retinues".[84]

Historians who wrote about the Mamluk period could be classified into three groups. The first group are local historians who witnessed the historical events or were alive at some stage during that period, such as al-Suwaidi, Fa'iq and al-Karkukaly. This group usually follow the traditional methods of storytelling and focus on the rulers more than other measures in society, and they tend to criticise negative issues gently to maintain good relations with the governors. The second group are the foreign visitors who recorded their impressions, such as Niebuhr, Buckingham and Longrigg. The focus of their writings is on political issues, tribal conflicts and the relations between different powers. The third group includes modern local historians like Nawras and Izz al-Din, who refer to the writings of the first and second groups in their historical narrations, which makes these references an interesting mix of transmitted and innovative ideas.

The overall picture of Mamluk rule in Baghdad suggests five main characteristics of this period: firstly, promoting the idea of reforms through the unfair treatment of land usage and inconsistency with building and renovating. Secondly, a tendency for more independence from Istanbul. During this period, Iraq witnessed many attempts to establish strong local governments with more enhancement of the idea of uniting Iraq under the sovereignty of Baghdad as a central city.[85] Thirdly, a great determination to govern

Baghdad. The position of a governor of Baghdad was an outstanding goal for many people who were keen to achieve it. Izz al-Din notes the *wālī* who governs Baghdad was considered the best, so people would pay money and do whatever they could to get this position.[86] Fourthly, there was continuous trouble in relation to tax collection, and fifthly, warm relations with Europe, especially the British.

The examination of the history of the Mamluks in Baghdad highlighted the conditions of Baghdad at the time of the arrival of many emissaries. This would assist in understanding the circumstances from different angles. In addition, it showed the complexity of history that may encompass both factual and non-factual thoughts. Therefore, to better understand history it is crucial to examine multiple sources rather than focussing on a particular one. The historiography of 18th and 19th-century Baghdad indicates a typical situation of diverse attitudes and techniques to portray historical events. The interpretation of events depended on the aims of the writing and temporary impression of the author. This situation points out the coexistence of contradictory issues in historiography, and urges the need for more awareness in dealing with the writing of history.

Notes

1 Among the evidence of its old age is that it was mentioned in documents of King Hammurabi in 1750 BC. Also, the Aramaic word 'bakdādu' was engraved on the old Babylonian mud stone that was discovered in 1780. See al-Warid, BA 1980, *ḥawādith Baghdad fī 12 qarn* (Arabic), al-dār al-'arabiyya, Baghdad, p. 228.

2 The Abbasids state lasted between 750 and 1258.

3 The age of the city is counted from the year 767 when the construction concluded, until 836 when they moved the capital to Samarra.

4 Cooperson, M 1996, 'Baghdad in rhetoric and narrative', *Muqarnas: an annual on the visual culture of the Islamic world. Volume 13, Aga Khan Program for Islamic architecture*, pp. 99–113. Despite great excavation efforts, the Department of Antiquities in Baghdad could not find any trace of it. See Ra'uf, IA 2000, *ma 'ālim Baghdad fī al-qurūn al-muta'khira* (Arabic), Bayt al-ḥikma, Baghdad, Iraq.

5 Lassner, J 1970, *The topography of Baghdad in the early Middle Ages: text and studies*, Wayne State University Press, Detroit, MI, p. 27.

6 The settlement in the eastern part existed thousands of years before *al-karkh* settlement, yet it was expanded during that period.

7 See Ra'uf (ed.) 2008, *'akhbār Baghdad wa mā jāwarahā min al-bilād*, by al-Alusi M Sh (Arabic), al-dār al-arabiyya lil mawsū'āt, Beirut, p. 58. Also see Selman, I, 'abd al-Khaliq, H, al-Izzi, N & Yunus, N 1982, *al- 'imārāt al- 'arabiyya al-Islāmiyya fī al-Iraq* (Arabic), vols. 1&2, al-Hurriyya Press, Baghdad.

8 For more details about the history of Baghdad see Jawad, M, Susa, A, Makkiya, M & Ma'ruf, N 1968, *Baghdad* (Arabic), Iraqi Engineers Association with

Gulbenkian Foundation, Baghdad. Also see Al-Warid, *ḥawādith Baghdad fi 12 qarn*.

9 For more information see Choueiri, YM 2005, *A companion to the history of the Middle East*, Blackwell companions to world history, Blackwell Pub. Ltd, Malden, MA. Also see Khayyat, J (ed.) 1968, *'arba' at qurūn min tārīkh al-Iraq al-ḥadīth*, by Stephen Hemsley Longrigg (Arabic), matba'at al-tafayyud. Also see Choueiri, *A companion to the history of the Middle East*. And see Makkiyya, *Baghdad*.

10 Morkoc, SB 2010, *A study of Ottoman narratives on architecture: text, context and hermeneutics*, Academia Press, Bethesda, MD, p. 144.

11 Raymond, R 2005, 'Urban life and Middle Eastern cities: the traditional Arab city', in Choueiri, YM (ed.), *A companion to the history of the Middle East*, Blackwell companions to world history, Blackwell Pub. Ltd, Malden, MA, p. 207.

12 Bonine, ME 2005, 'Islamic urbanism, urbanites, and the Middle Eastern city', in Choueiri, *A companion to the history of the Middle East*, p. 403.

13 Grabar, O 1987, *The formation of Islamic art*, Rev. and enl. edn, Yale University Press, New Haven, CT, p. 70.

14 Moholy-Nagy, S 1968, *Matrix of man: an illustrated history of urban environment*, Pall Mall Press, London, p. 35.

15 Moholy-Nagy, *Matrix of man*, p. 81. Also see Oleg Grabar's remarks about the plan of Baghdad in the Aga Khan Award for Architecture 1986, *Architecture education in the Islamic world: proceedings of seminar ten*, Granada, Spain, Concept Media Pte. Ltd., Singapore, p. 35.

16 See Ra'uf (ed.) *'akhbār Baghdad wa mā jāwarahā min al-bilād*.

17 For more details refer to Chisholm, H 1911, *The encyclopaedia Britannica*, Vol. 7, Constantinople, the capital of the Turkish Empire.

18 Morkoc, *A study of Ottoman narratives on architecture*, p. 127.

19 The narration of Abadah about khan Lāwand which was a residency for soldiers is evidence. He narrates if Istanbul is pleased with someone he will never be punished on any crime he commits. See Ra'uf (ed.) 2004, *al-'iqd al-lāmi' bi-'āthār Baghdad wal-masājid wal-jawāmi'* (Arabic), *by 'abd al-Hamid Abadah*, First edn, 'anwār dijla Publishing, Baghdad, p. 202.

20 See al-Warid, *hawādith Baghdad fi 12 qarn*. Also see Choueiri, *A companion to the history of the Middle East*.

21 Said, EW 2007, *On late style: music and literature against the grain*, 1st Vintage Books edn, Vintage Books, New York, NY, p. 35.

22 Hamadeh, SH 2007, 'Public spaces and the garden culture of Istanbul in the 18th century', in Aksan, VH & Goffman, D (eds), *The early modern Ottomans: remapping the empire*, Cambridge University Press, Cambridge, pp. 299-310.

23 An example of these studies: Atasoy, N 2004, 'Ottoman garden pavilions and tents', in *Muqarnas*, vol. 21, *Essays in honor of J. M. Rogers* pp. 15–19.

24 Veinstein, G 2008, 'The Ottoman town; fifteenth-eighteenth centuries', in Jayyusi, SK, Holod, R, Petruccioli, A & Raymond, A (eds), *The city in the Islamic world*, Brill, Leiden, Boston, MA, pp. 207–212.

25 Veinstein, 'The Ottoman town', pp. 207–212.

26 Veinstein, 'The Ottoman town', pp. 207–212.

27 Veinstein, 'The Ottoman town', pp. 207–212.

28 Inalcik & Quataert, D 1997, *An economic and social history of the Ottoman Empire, volume 2: 1600–1914*, 2 vols, Cambridge University Press, Cambridge,

p. 646. For more information about the Ottoman town see Kafesci oglu, C 2009, *Constantinopolis/Istanbul: cultural encounter, imperial vision, and the construction of the Ottoman capital*, The Pennsylvania State University Press, University Park, PA. Also see Cerasi, M 2005, 'The urban and architectural evolution of the Istanbul divanyolu: urban aesthetics and ideology in Ottoman town building', *Muqarnas*, vol. 22, pp. 189–232.

29 Inalcik & Quataert, *An economic and social history of the Ottoman Empire*, p. 673.

30 Tripp, C 2007, *A history of Iraq*, 3rd edn, Cambridge University Press, Cambridge.

31 Abdullah, T 2001, *Merchants, Mamluks, and murder: the political economy of trade in 18th-century Basra*, SUNY series in the social and economic history of the Middle East, State University of New York Press, Albany, NY, p. 6.

32 Veinstein, 'The Ottoman town', pp. 207–212.

33 Veinstein, 'The Ottoman town', pp. 207–212.

34 Lafi, N 2007, 'The Ottoman municipal reforms between old regime and modernity: towards a new interpretive paradigm', in Cihangir, E 2007(ed.), *Uluslararas Eminonu Sempozyumu: tebligler kitab International Symposium on Eminonu: the book of notifications*, Eminonu Belediyesi Baskanlg, Istanbul, p. 354.

35 Lafi, 'The Ottoman municipal reforms between old regime and modernity', p. 355.

36 Lafi, 'The Ottoman municipal reforms between old regime and modernity', p. 356.

37 Veinstein, 'The Ottoman town', pp. 207–212.

38 Nawras, AMK 1975, *hukkām al-mamālik 1750–1831* (Arabic), silsilat al-kutub al-haditha 84, al-maktaba al-wataniyya (the National Library) number 611, Baghdad.

39 This book is also called *mir'āt al-zawrā'*. It was written in Turkish. The author Sulayman Fa'iq is the son of Talib agha who had a high position during the ruling of the last Mamluk governor Dawud *pāshā* (1817–1831).

40 Al-Warid, *hawadeth Baghdad fi 12 qarn*, p. 200. See also Jawad, M & Susa, A 1958, *dalil khāriṭat Baghdad al-mufaṣṣal fi khiṭaṭ Baghdad qadiman wa ḥadīthan* (Arabic), al-majma' al-'ilmi al-Iraqi, Baghdad, p. 292.

41 Longrigg, SH 1968, *Four centuries of modern Iraq*, Librairie du Liban, Beirut, pp. 81–82.

42 Those taxes are the Badj (الباج) and Tamgha (الطمغة). The Badj is a tax on everything entering Baghdad by land and Tamgha is a tax on things coming to the city by river. See Khulusi, S (ed.) 1962, *tārīkh Baghdad:hadiqat al-Zawrā' fi sirat al-wuzarā' by 'abd al-Rahman al-Suwaidi* (Arabic), vol. 1, maṭba'at al-Za'im, Baghdad, p. 68.

43 Khulusi, S (ed.), *tārīkh Baghdad*, pp. 18–32.

44 Nawras, *hukkām al-mamālik 1750–1831*, p. 25.

45 Abdullah, *Merchants, Mamluks, and murder*, p. 11.

46 Al-Warid, *ḥawādith Baghdad fi 12 qarn*, pp. 217–219.

47 Khayyat (ed.), *'arba'at qurūn min tārīkh al-Iraq al-ḥadīth*, p. 155.

48 Khayyat (ed.), *'arba'at qurūn min tārīkh al-Iraq al-ḥadīth*, pp. 154–155.

49 Longrigg, *Four centuries of modern Iraq*, p. 11.

50 Mohammed Ali, IM 2008, *Madinat Baghdad: al-'ab'ād al-ijtimā'iyya wa ẓurūf al-nash'a* (Arabic), al-haḍāriyya lil ṭibā'a wal nashr, al-'ārif lil maṭbū'āt, Baghdad, p. 139.

51 The corresponding Arabic term for discipline is تأديب.
52 Fa'iq beg, S 2010, *tārīkh Baghdad* (Arabic), 1st edn, dār al-rāfidayn for publishing, Beirut.
53 Longrigg, *Four centuries of modern Iraq*, p. 170.
54 Fa'iq beg, *tārīkh Baghdad*, pp. 18–19.
55 Inalcik & Quataert, *An economic and social history of the Ottoman Empire*, p. 676.
56 This position was called katkhudā (كتخدا)
57 Fa'iq beg, *tārīkh Baghdad*, p. 19.
58 Fa'iq beg, *tārīkh Baghdad*, p. 20.
59 Ra'uf (ed.) 1978, *tārīkh ḥawādith Baghdad wa al-Basra 1186–1192 AH, 1772–1778 AD by 'abd a-Rahman al-Suwaidi* (Arabic), The Ministry of Education and Arts, Baghdad.
60 Fa'iq beg, *tārīkh Baghdad*.
61 See Fa'iq beg, *tārīkh Baghdad*. Also see Ra'uf (ed.), *tārīkh ḥawādith Baghdad wa al-Basra*. Also see Ra'uf, *al-'iqd al-lāmi' bi-'āthār Baghdad wal-masājid wal-jawāmi'*.
62 Nawras, *hukkām al-mamālik 1750–1831*, p. 44. See also al-Karkukaly, R 1992, *dawḥat al-wuzarā' fi tārīkh ḥawādith Baghdad al-Zawrā'* (Arabic), al-Sharif al-Raḍi Publishing, Qum.
63 According to Abdullah, both Sulayman abu Layla and Sulayman al-Kabir were *mutasallims* (rulers) of Basra before their accession to the post of *wali* of Baghdad and the same is true with Ahmed *pāsha*, which explains their strong relations with the British Resident in Basra. See Abdullah, *Merchants, Mamluks, and murder*, p. 29.
64 Fa'iq beg, *tārīkh Baghdad*, p. 23.
65 Nawras, *hukkām al-mamālik 1750–1831*, p. 44.
66 Nawras, *hukkām al-mamālik 1750–1831*, p. 46.
67 Fa'iq beg, *tārīkh Baghdad*, p. 35.
68 Longrigg, *Four centuries of modern Iraq*, p. 198.
69 Longrigg, *Four centuries of modern Iraq*, p. 324.
70 Longrigg, *Four centuries of modern Iraq*, p. 257.
71 Longrigg, *Four centuries of modern Iraq*, p. 254.
72 For more details refer to al-Warid, *ḥawādith Baghdad fi 12 qarn*. Also see Nawras, *hukkām al-mamālik 1750–1831*.
73 Izz al-Din, Y 1976, *Dawud pāshā wa nihāyat al-mamālik fi al-Iraq* (Arabic), maṭba'at al-sha'ab, the University of Baghdad, p. 43.
74 Al-Warid, *ḥawādith Baghdad fi 12 qarn*, p. 239.
75 Al-Warid, *ḥawādith Baghdad fi 12 qarn*, p. 232.
76 Al-Warid, *ḥawādith Baghdad fi 12 qarn*, pp. 236, 239.
77 Al-Warid, *ḥawādith Baghdad fi 12 qarn*, p. 237.
78 Izz al-Din, *Dawud pāshā wa nihāyat al-mamālik fi al-Iraq*. Also see Nawras, *hukkām al-mamālik 1750–1831*.
79 Longrigg, *Four centuries of modern Iraq*, p. 250.
80 Khoury, DR 2007, 'Who is a true Muslim? Exclusion and inclusion among polemicists of reform in 19th-century Baghdad', in Aksan, VH & Goffman, D (eds), *The early modern Ottomans*, p. 273.
81 Longrigg, *Four centuries of modern Iraq*, p. 30.
82 Inalcik & Quataert, *An economic and social history of the Ottoman Empire*, p. 761.

83 Nawras, *hukkām al-mamālik 1750–1831.*
84 Inalcik & Quataert, *An economic and social history of the Ottoman Empire,* p. 674.
85 Ra'uf (ed.), *tārīkh ḥawādith Baghdad wa al-Basra,* p. 5.
86 Izz al-Din, *Dawud pāshā wa nihāyat al-mamālik,* pp. 12–20.

3 Travelogues and History Representation

Travelogues' Methods and Attitudes

In order to attain a comprehensive understanding of the role of travelogues in representing the urban history of Baghdad, it is important to highlight their methods and attitudes of representation, to evaluate their position as a source in Baghdad's historiography, along with other historical sources. As outlined before, travel writings of Baghdad were written by two groups; regional travellers and European travellers. The writings of each group were greatly influenced by their individual experience and their means of understanding the city. Subsequently, travel writing in Baghdad experienced a dramatic change in modes through different periods of its history. In the early centuries after the foundation of Baghdad, travellers' texts were inspirational, emotional and appealing. By the 15th century, the attitudes started to take a political overtone, influenced by the writer's aspirations. In the 18th century, regional writers expressed nostalgic modes, while European writers indicated mixed modes of sighting with political curiosity.

Typically, travel writings place a great emphasis on the visual quality of a place, and on its generic appearance. The traveller usually approaches the city with particular conceptions in his mind. This stage can be considered as the stage of 'initial recognition'. When the traveller reaches the city, the second stage begins, as he visually perceives the city, his imaginative preconceptions become rationalised. While some of these imaginative pictures improve, others will be reduced or vanish completely, depending on the circumstances of the city. This stage can be outlined as the stage of 'visual recognition'. The writing of travelogues in general is the third stage of this framework of comprehension, when the ideas are combined and presented in a specific way that reflects the traveller's ultimate understanding. This stage can be outlined as the stage of 'reflective recognition'. These steps show the crucial impact of each individual's experience and intentions on the context of his writing.

DOI: 10.4324/b23141-4

In contrast to the theme of passionate expression that is exposed in poetry and literature, discovery usually distinguishes travelogues from other types of historical writing. Discovery involves observing, revealing, finding and interpreting, in addition to disclosing facts and documents. Historically, the theme of discovery took a broader meaning, when it attempted to organise different interpretations of specific issues in society, such as sociability and entertainment. Chard classifies travel writing at the end of the 18th century into two categories; the romantic view of travel and the touristic view.[1] Within the romantic view, travel is considered a form of personal adventure that holds out a realisation of self through the exploration of the other. It also entails crossing symbolic as well as geographical boundaries and inviting different forms of danger and destabilisation.

The second approach is the view of the tourist "who recognises that travel might constitute a form of personal adventure, and might entail danger and destabilisation, but … attempts to keep the more dangerous and destabilising aspects of the encounter with the foreign at bay".[2] Although she identifies them as opposing observations, these two categories seem to be different versions of the same entity, because both approaches involve discovery and entail danger on a certain level. In their formal, compositional and symbolic characteristics, the forms of the city convey a remarkable phenomenon in the representation of texts. This phenomenon is at once simple and complex. It is simple in that its physical character can be identified in various conventional studies that elaborate material qualities. Yet it is complex in that the representation suggested by textual interpretation has little to do with formal criteria and has unlimited capacity for meaning. The interpretation by travelogues presents a number of ideas that reflect discrete versions of the past. However, some of these ideas act as catalysts for insights into all groups of visitors. For example, the Tigris River gained the appreciation of both Baghdadi scholars and various travellers, though its major influence and attachment was on the residents and regional visitors.

In regard to Baghdad specifically, the degree of familiarity with and awareness of particular matters influenced the presentation of Baghdad by different visitors. The poets and regional scholars of Baghdad were familiar with the place, language and all other aspects of the society. Whereas the emissaries needed to apprehend situations that are unfamiliar to them, and they "were likely to encounter many adventures from which they could extricate themselves only through skill and daring".[3] This approach usually entails problems in understanding the society in discovery, and therefore acquires additional efforts to learn the rules and norms of that society. In addition to touristic views, travel writings of Baghdad in the 18th and 19th centuries show purposeful and self-oriented approaches that require significant analysis of the place. Individual adventures in these writings were devoted to present the 'other' through personal judgements.

Since the emissaries often experienced formal situations, their writings indicated limited romantic intentions and incorporated fewer emotional expressions. These people were keen to protect themselves from unexpected situations experienced with some other trips at that time, and they were determined to complete their tasks quickly and leave the place. They recorded things that they were able to physically observe in their visits. Hence, this kind of literature can be located in the domains of topography and geography rather than social or cultural history. Regardless of their own specialities and objectives, they were often interested in exploring the history of Baghdad, which complicated the connotation of historical matters. Since history is a multifaceted discipline that contains transmitted knowledge, the writing of history in these travelogues became an embodiment of some kind of invention. There is always a fine line between 'invention' and 'discovery', which makes it difficult to distinguish each phenomenon separately.

The 18th century may have been the initial stage of the invention of Baghdad's history. Combined themes of discovery and invention became the trait of successive writers since the ambiguity of history allowed them to suggest interpretations for specific events. Unfortunately, these interpretations stimulated more inventions in historical studies. In addition to the lack of reliable historical resources in the travel writings of the 18th and 19th centuries, myths of backwardness and primitiveness affected the emissaries' thinking, and influenced their writings greatly. The merging of different ideas resulted in them being transferred from their original status to a collection of obscurities. Despite the mixed ideas offered by these writings, conventional studies consider the travelogues of the 18th and 19th centuries as favourable references, because of the plentiful information they provide and the shortage of other writings of that period. However, they could be considered among the multifaceted historiographical references that interconnect various perspectives of Baghdad from inside and from outside, to ensure an integrated approach to the understanding of history.

The travelogues of the 18th and 19th centuries focused on the material forms of Baghdad, which were sometimes a disappointment for these visitors, because they approached the city with highly romantic preconceptions inspired by tales from the Abbasids' time. These emissaries "were expected to confirm the image of the Middle East purveyed by the romantic literature and the art of the time - that of a sensual, dangerous and above all different place".[4] Associating history with a particular feature of ostentation results in unrealistic accounts that do not consider change as a common historical attribute. The emissaries of Baghdad in this period enjoyed free access to the exterior forms of the city, so they reflected a lot on the natural environment and specific architecture. These components presented the main characteristics of the city for them.

The significance of Baghdad as a hub for advanced learning, and as a marketplace is overlooked in these texts. In addition, the meaning of some urban forms that were used as venues for learning, such as mosques and schools, was associated with material qualities rather than educational, spiritual and social values meanings. While many travelogues of the 18th and 19th centuries opened up the way for misunderstanding, similar approaches of subsequent historical writings contributed to the historiographical problem as well. The continuous transmission of, and additions to, historical contents and the divergence of historical ideas, are considered as a transgressive act that is 'a postmodern gaming or poaching'[5] Nonetheless, considering travelogues among the literary sources of Baghdad's historiography opens up opportunities to observe more characteristics of Baghdad.

Therefore, all kinds of travel writing take an important role in representing the urban history of Baghdad. The explorers' interpretation considers more social and environmental aspects, in addition to the visual aspects, which makes it a basic source of social, educational and architectural history. On the other hand, the emissaries' interpretation depends hugely on the observation of the outer appearance, which makes them adequate in the mapping and geographical fields. However, sometimes both resources provide additional information in relation to similar fields. In short, the travelogues of Baghdad embrace both interpretative and unreliable ideas, compared to other writings of the past. The remarkable achievement of these texts is the provision of a perception of Baghdad from outside. These insights complement sensitive meanings from inside to develop a better understanding of the history of this city.

Travelogues and Orientalism

In the context of studying the travelogues of the 18th and 19th centuries, it is essential to discuss the idea of 'Orientalism' because of its strong relevance to these texts that contributed to the wide spread of information about the places they visited at that time. In his book; *Orientalism*, Said notes:

> There were of course innumerable voyages of discovery; there were contacts through trade and war. But more than this, since the middle of the 18th century there had been two principal elements in the relation between East and West. One was a growing systematic knowledge in Europe about the Orient, knowledge reinforced by the colonial encounter as well as by the widespread interest in the alien and unusual, exploited by the developing sciences of ethnology,

comparative anatomy, philology, and history; furthermore, to this systematic knowledge was added a sizable body of literature produced by novelists, poets, translators, and gifted travellers.[6]

European travelogues, along with other forms of writing, established a system of knowledge about the Orient. The word 'Orient' was utilised to differentiate between the countries to the east on planet Earth, mainly Islamic countries, from the countries to the west on planet Earth, or the Occident, that are mainly European and Christian countries. This idea of differentiation between east and west is not a historical reality, it is rather a European invention. The Orient, in this context "is an integral of European material civilization and culture".[7] Orientalism represents the east culturally and ideologically as a mode of discourse with supporting institutions, vocabulary, scholarship, imagery, doctrines, even colonial bureaucracies and colonial styles.[8] Subsequently, this idea has been accepted by a large group of writers:

> among whom are poets, novelists, philosophers, political theorists, economists, and imperial administrators, have accepted the basic distinction between East and West as the starting point for elaborate theories, epics, novels, epics, social descriptions and political accounts concerning the Orient, its people, customs, mind, destiny, and so on.[9]

Following the invention of the Orient, a group of people who were concerned with the issues of Orientalism has emerged. Said defines an Orientalist as "anyone who teaches, writes about, or researches the Orient … and this applies whether the person is an anthropologist, sociologist, historian, or philologist … either in its specific or its general aspects".[10] According to this definition, the emissaries are considered Orientalists, since they write about the Orient and its particular aspects as "a hotbed of sensuality".[11] For them, the Orient constitutes:

> An analysable formation … for example, that of philological studies, of anthologies of extracts from Oriental literature, of travel books, of Oriental fantasies-whose presence in time, in discourse, in institutions (schools, libraries, foreign services) gives it strength and authority.[12]

Orientalism is not limited to individuals who are interested in the Orient. It incorporates any group or institution that is concerned with such matters. Although the term 'Orientalism' seems to belong to the pre-modern era, its implications have been transferred to be part of what is called 'Middle Eastern studies' or 'Arab and Islamic studies'. So, its material and ideas remained the same, but the term has been substituted with other names, which asserts the need to further comprehensive analyses of this ideology in a scientific, practical way. Said explains:

> For the Orientalists, there is a support system of staggering power, considering the ephemerality of the myths that Orientalism propagates. This system now culminates in the very institutions of the state. To write about the Arab Oriental world, therefore, is to write with the authority of a nation, and not with the affirmation of a strident ideology but with the unquestioning certainty of absolute truth backed by absolute force.[13]

Said implies writing about the Arab and Islamic world entails authority as the key motive for these studies. Despite the intense writings by the Orientalists, and their constant intrusion, they "remained outside the Orient",[14] which means they had no intention of understanding these regions truthfully. Due to their colonial drives, the imaginative Orient suggested by them "is taught, researched, administered, and pronounced upon in certain discrete ways",[15] to outline "a place of romance, exotic beings, haunting memories and landscapes, remarkable experiences".[16] Thus, because of the complex intentions of Orientalism, the Orient has never been "a free subject of thought or action".[17] As a result, the Orient became "less a place than a *topos*, a set of references, a congeries of characteristics, that seems to have its origin in a quotation, or a fragment of a text, or a citation from someone's work on the Orient".[18] Historians suggest the late 18th century as a rough starting point for Orientalism "as the corporate institution for dealing with the Orient by making statements about it, authorizing views of it, describing it, by teaching it settling it, ruling over it".[19] Though the 18th century has been a body to inspect the Orient, the ideas, motives and first steps might have taken place in the 16th century, with the growing interest of the Europeans to intervene in these lands and benefit from their great heritage and civilisation.

As stated by Said, there were two principal methods by which Orientalism delivered the Orient to the West in the early 20th century; the first one by means of the administrative capacities of modern learning, the universities, the professional societies, geographical organisations and the publishing industry. The second method was the result of the merging of ideas, by speaking about the Orient, translating texts, explaining civilisations, religions, dynasties, cultures, mentalities as academic objects, screening off from Europe by virtue of their inimitable foreignness.[20] These key steps in Oriental scholarship were first implemented in Britain and France, then expanded by the Germans "yet what German Orientalism had in common with Anglo-French and later American Orientalism was a kind of intellectual authority over the Orient within Western culture".[21]

Therefore, Orientalism cannot be considered "an airy European fantasy about the Orient but a created body of theory and practice in which, for many rations, there has been a considerable material investment".[22] The real image of the Orient is that it was orientalised not only because it was discovered as Oriental "in all those ways considered common-place by an average 19th century European, but also because it could be … that is, submitted to being made Oriental".[23] Enforcing an idea and gaining acceptance with all willpower is one of the major psychological tricks of this scheme. Orientalism had remarkable implications for Islam, as its purpose was not so much to represent Islam itself but to represent it for the medieval Christian.[24]

The analyses of the roots of the concept of 'Islamic architecture' proves that it is beyond question that Orientalism is not a rational or ideological notion, but rather a political one. European travellers had a big role in these policies, "to write about Egypt, Syria, or Turkey, as much as travelling in them, was a matter of touring the realm of political will, political management, political definition".[25] Further, Orientalists often reiterated the theme that Islam was oppressive, and it "should be severely criticized for its deviation from moral standards".[26] According to Said, 20th-century Orientalism has "successfully controlled freedom and knowledge".[27] He notes that if this knowledge has any meaning, then it will be "a reminder of the seductive degradation of knowledge, of any knowledge, anywhere, at any time".[28] Thus, it becomes clear Orientalism has been a Western style of dominating, restructuring and having authority over the Orient.[29] Black's statement about Baghdad sums up the aims, attitudes and principles of this doctrine:

> Civilization was once Mesopotamia's greatest export. But the balance of payments was tragic. Little came back. The land between the two rivers entered the 19th century so far behind, so bereft of the gifts it had bestowed upon the world that the advanced nations to the west would not respect it as anything more than a domain ripe for domination.[30]

This exaggerated statement on Baghdad[31] confirms the effects of Orientalism on the way of thinking of those writers. It seems that imagination and invention were optimal in order to establish a foundation for the Europeans to colonise Islamic countries. Nevertheless, as outlined earlier, the terms Orientalism and Oriental studies are disappearing, and they are less preferred by specialists nowadays, because they are too vague and general, and they connote "the high-handed executive attitude of nineteenth-century and early twentieth-century European colonialism".[32] New organisations replaced Oriental institutes, and held different titles due to current requirements, but the core of the new studies is still strongly dependent on concepts of Orientalism.[33]

Travelogues and Islamic Studies

The investigation of travelogues in relation to 'Islamic studies' is crucial to this book, since Baghdad is among the big cities of the Islamic world. In order to understand the motives of travelogues, it is necessary to recognise the basis of concepts like Islamic city and Islamic architecture, because these ideas affected the history of Baghdad significantly. Raymond suggests the principal creators of the concept of the 'Islamic city' and its body of research were a number of French Orientalists from the 1920s who proposed that as a doctrine, judicial system and more generally as a culture, Islam determines a specific city model, which they called 'Muslim town'.[34] He considers Grunebaum's article 'the structure of the Muslim town',[35] as the account of the final formulation of the notion of the Islamic city, and an effective text that sums up this doctrine[36] which highlights a direct connection between the Islamic city doctrine and Orientalists' perspectives.

Associating a city with religion resulted in continuous major critiques of the Islamic city in literature. Studies of what are called 'Muslim towns' or 'Arabic towns' developed contradictory senses of admiration, degradation, remoteness and unfamiliarity. The association of these cities with a definitive religion, nationality and geography restricts the historical experience and generates a feeling of a unique, yet ambivalent, history. Raymond notes these texts deliver negative conclusions on traditional towns because they "clearly display the influence of a colonial spirit".[37] The superficial understanding of Islam in the Orientalists' minds, and their preoccupied judgements of Muslim cities, greatly influenced the methodology of writing about those cities. In addition, the use of specific linguistic terms affected the way of interpreting historical events. Although Islamic city doctrine invoked a great amount of research, it added much misperception to the historical research on Islamic cities. The interest in Islamic studies through this doctrine came to a height in the 1970s. This culmination resulted in the founding of many specialised organisations and the gaining of acceptance globally. The aim of establishing those organisations was to promote the study of Islamic art, architecture, urbanism, landscape design and conservation, to apply that knowledge to contemporary design projects.

Although some of these organisations were not necessarily connected directly to the Orientalists' doctrine of the Islamic city, they were intellectually influenced by its concepts. The most significant example of these organisations is the Aga Khan Development Network, which established the Aga Khan Programme for Islamic Architecture in 1979 at Harvard University. This organisation also established the Aga Khan Award for Architecture in 1977 and the Aga Khan Trust for Culture in 1988. Also, at Massachusetts Institute of Technology in 1983, the journal *Muqarnas* was founded. Another example of these organisations is the Centre for Middle Eastern Studies that was established in the 1970s in Rutgers University. And a third example is the *Journal of the Islamic Environmental Design Research Centre* founded in 1985 by the Italian architect Attilio Petruccioli. These organisations continued to adhere to the idea of the Islamic city model and Islamic urbanism. The basic concept launched and enhanced by those institutions is that there 'was' an Islamic city that constituted 'Islamic architecture' and this architecture had special qualities. Consequently, some terms, such as culture, heritage, identity of place, traditional, Arabic cities and Middle Eastern, became key assets for the preservation process and the main focus of those studies.

This movement received international acceptance of the idea of studying architecture in relation to 'other' place and 'other' history. Educational institutions in the Islamic world in the 1980s, thinking they had productive effects on society, started to implement Orientalist ideas. Subsequently, the topic of heritage emerged in architectural education programmes as a great guide for conservation, and scholars and architects of that period were keen to uphold the new proposal. As a result, extensive bodies of research were initiated to study topics like the guidelines for conserving the built environment, institutions of Muslim society, traditional Muslim philosophy of life, the role of Islamic laws, preservation plans and innovative ideas of morphological structuring principles.

Books by Besim Hakim in 1986 and Jamal Akbar in 1984 considered some of these topics.[38] This approach instigated the concept of heritage maintenance that focused on Islamic art, calligraphy, water systems and gardens more than conceptual frameworks and experience. However, Bonine suggests that in the late 1980s another idea was initiated to examine the impact of Islam on the built forms; that is "to explore the inner motivations behind visual structures as the main source of pre formal shaping forces".[39] Although preservation is the outline aspect of such studies, the danger lies in labelling the outstanding work of the past as 'heritage' which applies that they are no longer suitable to be used or replicated in modern society. An additional field of study that reflects on social urban history has emerged later, which reflects the mounting interest in social sciences to understand these societies. This field focusses on the unique architectural heritage of Islamic cities, calling it 'Islamic architecture'. In many cases, associating architecture with a religion, such as Islam, requires evidence.

In the process of analysing a Muslim town, the study of the urban history of a place would provide clues on the articulation of the space. This involves the study of many aspects, including architectural, cultural and political history in addition to aesthetic aspects. There is no doubt that the similarities between architectural components of different cities in the Islamic world reflect some ideological and performance comparisons. Also, it can be assumed that Islamic principles affected the articulation of cities and the construction of architectural forms. Although Islam made people aware of nature, and enriched human culture, thought and philosophy, Islamic rules do not contradict scientific development. So, Islam may express the theory of architecture and its guidelines rather than architectural science. Accordingly, the identity of urban heritage in Islamic cities relates to a number of other factors in addition to religion. Therefore, connecting architectural heritage with Islam may be problematic.

It appears that the notion of 'Islamic city' is a culmination of colonialist and secularist complications in the 18th and 19th centuries. The secular approach has changed architectural and urbanism thinking in Europe, and consequently influenced colonised areas. Ansari suggests the basic thoughts of the secular movement are that religious authority will not be allowed to interfere in the secular sphere of life, and that time is tangible, and the march of history implies an irreversible process.[40] This meant, according to secularism, religion should have a marginal place in the communities that implement this idea and relating to religion is a sign of backwardness. In the region of Baghdad, the concept of secularism informed many architectural procedures, as several Iraqi architects were educated in Europe in the early 20th century. Some scholars consider it a "natural development of the twentieth century".[41] On the other hand, Pamir suggests that the changes in architectural attitudes in Istanbul and in other major cities took place effectively after developing the *tanzimat* in 1839.[42] Subsequently, architectural plans were provided either by foreigners coming from abroad or by locals who had architectural training in Europe, and this was heavily reflected on the city structure and forms.

The Islamic city's enquiry presents three approaches: material documentation, spatial analysis and segregation and isolation. The focus of the European travelogues on the physical components of cities established the base for the material documentation approach. Subsequently, a number of architects and art historians who are specialised in Islamic studies, such as Bianca and Grabar, emphasise the role of documenting architectural forms as identity providers and as a tool for historical lucidity. They assert that substantial historical figures encompass unique meaning within their formal components; hence they should be carefully preserved. Those scholars oppose the idea of privileging textual representation over physical documentation.

Likewise, Ousterhout proposes there is a remarkable change in focus from physical forms to ideas in the study of architectural and urban history. That is because current historians are "too bored with the actual bricks and mortar to read a technical report or a primary source, instead they focus their discourses on the history of scholarship and on the apocrypha of history".[43] He argues that sometimes bricks and mortar are not ideological signifiers, though "they can inform us about cultural transformations in a somewhat different way".[44] He recommends that more fieldwork is necessary to document the vanishing heritage in the eastern Mediterranean, and "this, not interpretation, must be the first task of the architectural historian".[45]

Undoubtedly the accumulated wealth of documentation of architectural forms enhances appreciation of the architectural experience and provides symbolic analyses of tangible characteristics. Yet, documentation alone cannot put forward a deeper interpretive language, since "it's never enough for the good building process to focus on the outside frame and to choose this frame in an abstract way separated from reality".[46] Perhaps the solution to this problem would be not to privilege documentation over the representation of ideas, but to consider it as a parallel source of information and an aid for more understanding of the historical experience.

The second method in the interpretation of Islamic cities is the focus on spatial analysis rather than the mass of buildings. Historians such as Edmond Bacon, Sibel Moholy-Nagy, Christine Boyer and others have strongly supported this method in their writings. Bacon suggests studying the nature of society, circumstances and the diversity of decisions made by people who live in it. He notes a "deeper understanding of the interactions of these decisions can give us the insight necessary to create noble cities in our own day".[47] The idea of reading architecture as a space rather than reading it as a mass is among the vital approaches for understanding historical experience. However, historians have been searching for more references that establish specific connections between the objects and their rationale for creation and continuity, which are not provided by the spatial inquiry.

In the last few decades, scholars like Hans-Georg Gadamer and Hayden White have suggested the consideration of discursive subjects like literature, history, sociology and psychology as the main sources for a more inclusive interpretation of architecture and urban history. Said relates these subjects to spatial analysis, citing them as 'systems of rarefaction' that comprise their symbolic, signifying, international and formal differences from other groups.[48] By utilising these topics, the representation of texts acts as a means by which spatial concepts are reduced to tangible images, so the strength and magnitude of textual representation is embodied in this two-way interaction between material criteria and initial ideas of the urban experience.

The significance of 'other' aspects of built forms has been the focus of architectural historians since the 1970s. Implementing openness to other subjects and embracing flexibility and multiple ways of relating, contributes to more productivity and sensibility in dealing with sensitive subjects like history. This process also provides a way into the realm of interpretation and presents a method which allows the coexistence of various meanings. So, the incorporation of all architectural criteria such as mass, space, art, history and science, would develop a complete vision of architecture instead of the unbalanced single-focus visions. Although these approaches are global, applying them to Islamic studies enriched these studies, due to the huge amount of inherited literature and travelogues from those periods.

The third approach of Islamic studies is segregation. These studies consider the narratives that are inherited from the Ottoman period difficult to understand because they came from a remote period. Sudden and immense change in attitudes between the 20th century and the 19th century established a sense of remoteness and isolation from the period prior to the 20th century. Despite the fact that this period is not that distant in reality, the slight unfamiliarity with inherited texts is a result of the large gaps generated by modern historiography, emotional effects of modernity and the impacts of the travelogues, which shaded the history of the Ottoman era in oldness and antiquity. This psychological remoteness influenced the techniques of conventional historiography and created barriers to the interpretation process.

For some historians, understanding the texts of the past is difficult for the contemporary mind because of the "temporal distance between the buildings and the texts".[49] However, the last few decades witnessed an increase in studies about Ottoman architecture, which reflects the growing interest of historians in connecting to that world. The interest in studying the travelogues of the 18th and 19th centuries has also been increasing, though these texts comprise old and unfamiliar terms as well. In fact, the real challenge is not in reading and understanding those texts, but in drawing out realistic experience among mixed and contradictory events. In any case, the fundamental insights of those narratives add to the definition and limits of space and confront the psychological issue of remoteness and isolation.

The idea of studying Islamic cities was behind the emergence of a huge genre of studies that are concerned with the 'identity of place'. Those cities suffered from a great loss in architectural originality due to the consequences of colonialism. Studies of identity of place are supported by different scholars because they are "strongly related to the cultural values of society",[50] and this is reflected on urban forms as a sense of place at multiple levels. It is vital for any nation to be aware of its origins to promote its successful progression. If the nation becomes less conscious about its origins, it becomes like an orphan who lacks the sense of belonging. Nevertheless, the 'identity of place' topic became a hub for some invented ideas, such as space, place, non-place, authenticity and culture.

While the prevailing aspects of these studies are usually the physical settings of heritage, in relation to architecture those concepts are discussed in relation to history, geography, art and sociology. The invention of the notion of culture as a main representative of local identity is one of the main products of the Islamic city project. The global endeavour to promote socio-cultural studies is reflected in the plentiful writing, which made cultural studies compatible to Islamic studies. Rapoport suggests that sociocultural variables play a major role in the urban environment, as they are a form of connection with the past and a medium for controlling communications among individuals and groups. He notes the built environment can usefully embody cultural ways of patterning,[51] which is a way of reducing the burden of social and environmental information.

However, Rapoport brings up a comparison between American cities and Islamic cities as exemplifiers of different organisations of the urban setting: "while U.S. cities maximise movement and accessibility, traditional Moslem cities limit movement and control behaviour by controlling mobility".[52] He associates American cities with geographical settings, whereas he associates Muslim cities with religion. This attitude makes the comparison invalid and intensifies differences, which expresses some shortage in these studies that have been implementing same methods by the travellers and Orientalists.

It seems that after about a whole century of initiating those studies, the will to discover more about Islamic cities is increasing. Richard Ettinghausen suggests it is only recently that historians began to chart different criteria of Islamic cities, including political, social, linguistic, religious and intellectual principles.[53] The basic criterion of utilising culture is to validate material qualities by non-material values and by meaningful use of historical structures. The definition of the term 'cultural heritage' that has emerged subsequently, broadened to include cultural assets that are associated with beliefs, artistic works, which "exhibit an important interchange of human values over a span of time or within a cultural area of the world".[54] As a result, cultural studies became a fundamental theme in many disciplines, including architecture.

Samer Akkach argues that the notion of culture was from the very beginning inevitably linked to both notions of 'difference' and 'place', and difference was seen through the lens of culture as a cultural difference that is necessarily placed in a definable geography.[55] European travelogues have been among the major resources that initiated and enhanced difference. Consequently, architectural studies envisioned culture as a bounding system that is rooted in a specific geography. This approach created problematic challenges for Islamic cities, such as isolation and lack of recognition. Akkach points out that this approach has created boundaries that have now unfolded. Historians and theorists are rethinking the meanings, validity and relevance of categories that were once outlined by conventional racial, religious, geographic or cultural references, such as Arab, Islamic, Middle Eastern, Asian, European and so on.[56] If these boundaries are created by current conventional definition of culture that took it away from its original meaning, then this term needs a definition that outlines its origins.

On the whole, the doctrine of the Islamic city was formulated to imply that the Islamic city is a non-city and Muslim urbanism is a non-urbanism. It refers to the Muslim town as nothing but the ruined image of a fine ancient order.[57] In addition, culture in general has been articulated to substitute Islam, and any actions related to specific geography or specific religion. However, if culture presented in its original meaning that entails civilisation, it would contribute to establishing positive diversity in a coherent world. Perhaps celebrating the achievements of the past positively, reforming the meaning of culture by introducing it as a form of human accomplishment, acknowledging the role of religion in forming the meaning of place and examining travelogues and other sources carefully and comparatively, would reduce the confusion in current cultural studies. In this way, the role of culture in inspecting collective memories, history writing, forming national identities and representing people's civilisation as a whole, becomes more informative, truthful and productive.

Travelogues and Heritage Understanding

The concepts of heritage and conservation are among the concepts that arose as part of the Islamic studies project. Baghdad's heritage that belongs to the 18th and 19th centuries contains a wealth of information that is yet to be discovered. In Arabic, the term '*turāth*' substitutes the English term 'heritage'. It usually refers to those inherited architectural and urban forms that represent unique aesthetic and structural features. The writings of the emissaries have influenced the understanding of heritage greatly, by informing the methods of conservation, disapproving some historical forms and enhancing modernisation. Despite the huge efforts that were initiated to sustain the architectural heritage in Baghdad by maintaining the physical forms, those figures continued to lose interactions with people, and they became obsolete with time.

Heritage forms are important indications of any nation's roots because they reflect its origins and provide connections between successive generations. Yet they are viewed to be unable to fully inform present and future architectural relations because of the fast and big changes that are happening in the world. Among the causes of the continuous destruction of heritage figures are the advancement of technologies, irresponsible political decisions, casual interpretation of history, unjust analysis of Islamic cities and dealing with heritage figures individually not collectively. For instance, recent research shows the supposed under-administration and non-administration of urban areas in the Ottoman period has been greatly exaggerated.[58] These attitudes led to a shortage of perfect strategies to maintain historical figures.

The major degradation of urban forms, and the spread of international language in architecture raised awareness of heritage among local architects. As a result, forums were held to discuss *turāth* protection, and different institutes were established to assist in this manner. In Baghdad and other cities, studies that aimed to maintain the Islamic heritage reached their peak in the 1980s. Iraqi architects like Ihsan Fethi, Muhammed Makkiyya and Rifa'at Chadirji led an extreme movement to appreciate traditional architecture. One of the key conservation schemes was the 'urban infill' project, which is supposed to repair the broken urban fabric by creating new typologies that relate to traditional models (Figure 3.1).

Figure 3.1 Proposed urban design sketch around Jami' Mirjan [Bianca 2000].

Other schemes include adapting local habits and climatic factors, in addition to implementing courtyards in new designs. Among the significant studies of heritage that intensified in the 1970s and 1980s to document the heritage of Iraqi cities that comprise what is called traditional architecture is the book *traditional houses in Baghdad*. This book studied Baghdadi houses in two sites: the city centre in *al-rusafa*, and *al-kazimiyya* city in *al-karkh* district,[59] and it was determined to document the wealth of specific Baghdadi houses that has never been documented before. The houses that were subjected to this study were inherited from 19th and early 20th-century Baghdad, though their style represented no particular phase in history. Chadirji suggests these architectural features are the accumulated stylistic expression of "four millennia of developments on the plains of Mesopotamia".[60]

Generally, these studies show a great sympathy with the continuous loss and damage suffered with time. Yet the solutions provided are not sufficient to conserve heritage because they superficially focus on the physical characteristics rather than the spirit, history and real meaning of place. Though they appreciate these buildings, they consider them 'historical', which implies they belong to history rather than the present time. The influence of the concepts of Orientalism is apparent on the way of thinking of these scholars, since most of them have been trained in the West, and therefore they transmitted these ideas to the field of heritage conservation.

Moreover, these studies ascribe considerable influence to Islam for the form and function of cities. According to the Orientalists' views, modernisation does not comply with religion. So, associating these forms with Islam entails rejecting their styles and keeping them in the museum of history. As outlined previously, Islam provides the guidelines for life and architecture, with no restrictions on specific structures. In his discussion of Islamic architecture, Chadirji questions the idea of religious architecture. He describes so-called Islamic architecture as 'a cultural form of expression', noting the term 'Islamic architecture' is inappropriate, for nobody speaks of Christian or Hindu architecture.[61] Unfortunately, European travelogues initiated many of these ideas.

Notes

1 Chard, *Pleasure and guilt on the grand tour*, p. 11.
2 Chard, *Pleasure and guilt on the grand tour*, p. 11.
3 Simpson, 'Arab and Islamic culture and connections', pp. 16–18.
4 Simpson, 'Arab and Islamic culture and connections', pp. 16–18.
5 Soder, H 2003, 'The return of cultural history; literary historiography from Nietzsche to Hayden White', *History of European Ideas*, vol. 29, no. 1, pp. 73–84.

6 Said, EW 1979, *Orientalism*, 25th anniversary edn, Vintage Books, New York, NY.
7 Said, *Orientalism*, p. 10.
8 Said, *Orientalism*, p. 10.
9 Said, *Orientalism*, pp. 10–11.
10 Said, *Orientalism*, p. 10.
11 Simpson, 'Arab and Islamic culture and connections', pp. 16–18.
12 Said, *Orientalism*, p. 28.
13 Said, *Orientalism*, p. 307.
14 Said, *Orientalism*, p. 222.
15 Said, *Orientalism*, p. 202.
16 Said, *Orientalism*, p. 9.
17 Said, *Orientalism*, p. 11.
18 Said, *Orientalism*, p. 177.
19 Said, *Orientalism*, p. 11.
20 Said, *Orientalism*, pp. 221–222.
21 Said, *Orientalism*, pp. 18, 19.
22 Said, *Orientalism*, p. 14.
23 Said, *Orientalism*, p. 13–14.
24 Said, *Orientalism*, p. 60.
25 Said, *Orientalism*, p. 169.
26 Simpson, 'Arab and Islamic culture and connections', pp. 16–18.
27 Said, *Orientalism*, p. 197.
28 Said, *Orientalism*, p. 328.
29 Said, *Orientalism*, p. 11.
30 Black, E 2004, *Banking on Baghdad,* John Wiley & Sons, Inc., Hoboken, NJ, p. 70.
31 Although it is true that 19th century Baghdad is less attractive than the tenth century Abbasid city, the literature composed by local scholars indicates the city was flourishing, and it witnessed many development schemes in the early 19th century.
32 Said, *Orientalism*, p. 10.
33 Said, *Orientalism*, p. 9.
34 Veinstein, 'The Ottoman town; fifteenth-eighteenth centuries', pp. 207–212.
35 This article was published in 1955. See Wilson, D (ed.) 1976, *Islam and medieval Hellenism: social and cultural perspectives*, Variorum reprints, London.
36 Raymond R, 'Urban life and Middle Eastern cities: the traditional Arab city', p. 207.
37 Raymond R, 'Urban life and Middle Eastern cities: the traditional Arab city', pp. 207–208.
38 Akbar, JA 1984, *Responsibility in the traditional Muslim built environment*, PhD thesis, Massachusetts Institute of Technology.
39 Bonine, 'Islamic urbanism, urbanites, and the Middle Eastern city', p. 394.
40 Ansari, MT (ed.) 2002, *Secularism, Islam, modernity: selected essays of 'Alam Khundmiri*, SAGE, London, p. 231.
41 Aga Khan Award for Architecture, *Architecture education in the Islamic world*, p. 22.
42 Aga Khan Award for Architecture, *Architecture education in the Islamic world*, p. 133.

43   Ousterhout, R, Necipoglu Gl & Aga Khan Program for Islamic Architecture (eds) 1995, 'Ethnic identity and cultural appropriation in early Ottoman architecture', in *Muqarnas: an annual on Islamic art and architecture*, vol. 12, E.J. Brill, Leiden, the Netherlands, pp. 48–62.

44   Ousterhout, 'Ethnic identity and cultural appropriation in early Ottoman architecture', pp. 48–62.

45   Ousterhout, 'Ethnic identity and cultural appropriation in early Ottoman architecture', pp. 48–62.

46   Al-Sadr, MB 2003, *al-Islam yaqūd al-hayāt, al-Madrasa al-Islāmiyya, risālatunā* (Arabic), Centre of special studies of Imam al-Sadr writings, Shariat, Qum, p. 180.

47   Bacon, EN 1967, *Design of cities*, Thames and Hudson, London, p. 13.

48   Said, EW 1985, *Beginnings: intention and method*, Columbia University Press, New York, NY, p. 300.

49   Morkoc, *A study of Ottoman narratives on architecture,* p. 18.

50   Oktay, B, Elwazani, S & al-Qawasmi, J (eds) 2008, *Responsibilities and opportunities in architectural conservation; theory, education, & practice*, vol. 2, CSAAR, Amman, p. 113.

51   Rapoport, A 1977, *Human aspects of urban form: towards a man-environment approach to urban form and design*, 1st edn, urban and regional planning series, v. 15, Pergamon Press, Oxford, pp. 265, 333.

52   Rapoport, *Human aspects of urban form*, p. 21.

53   Ettinghausen, R, Grabar, O & Jenkins, M 2001, *Islamic art and architecture 650–1250*, Yale University Press Pelican history of art, Yale University Press, New Haven, CT.

54   Oktay, *Responsibilities and opportunities in architectural conservation*, p. 113.

55   Akkach, S 2002, 'On culture', in Akkach, S & University of Adelaide, Centre for Asian & Middle Eastern Architecture (eds), *De-placing difference: architecture, culture and imaginative geography*, Centre for Asian and Middle Eastern Architecture, the University of Adelaide, p. 183.

56   Akkach, 'On culture', p. 184.

57   Raymond R, 'Urban life and Middle Eastern cities: the traditional Arab city', p. 220.

58   Raymond R, 'Urban life and Middle Eastern cities: the traditional Arab city', p. 220.

59   The Tigris River passes through Baghdad dividing it into two sectors; *al-karkh* in the western side of the city and *al-rusafa* in the eastern part. The administrative centre of the Ottomans is situated in *al-rusafa.*

60   See Warren, J & Fethi, I 1982, *Traditional houses in Baghdad*, Coach Publishing House, Horsham, England, p. 18.

61   Aga Khan Award for Architecture, *Architecture education in the Islamic world*, p. 22.

4 Diverse Travellers before the 18th Century

The observation of the travelogues of Baghdad before the 18th and 19th century periods is imperative, in order to recognise the types of travellers, their writing style, remarks and the various attitudes in their writings. Depending on their geographic location, the travellers of this period can be classified into two categories: regional travellers and European travellers.

Regional Travellers

Since the rise of Baghdad as a centre of learning and knowledge in the ninth century, travel literature has been an important part of Baghdad's history. When the Abbasids returned to Baghdad from Samarra'[1] in 278/892 they abandoned the round city that was their first planned city constructed on the western side of the Tigris River in 144/762.[2] Instead, they chose to reside in eastern Baghdad, which brought the east side of the city to a great long-flourishing era. With the opening of new schools, and the establishment of libraries and bookshops, scholars were keen to travel and stay in Baghdad for some time, due to the easy movement between Baghdad and adjacent cities. Many scholars wrote their travel observations in the form of historical books, as the subject of history received much attention at that time, along with other disciplines. Although these writings mainly focused on the events associated with the rulers rather than the urban structure of the city, they included remarkable indications of the architectural and urban criteria of Baghdad. The book *Baghdad*, written by ibn Tayfur in 280/893, is among the earliest books that present this kind of literature. Other historians who wrote about Baghdad are ibn al-Jawzi (d.597/1200), ibn al-Qufti (d.646/1248), al-Qazwini (d.682/1283) and ibn al-Futi (d. 723/1324).[3]

The most famous travelogue of Baghdad is the narrative of the Moroccan traveller, ibn Battuta. This 'explorer' visited Baghdad in 737/1337 on his way to Makkah, because Baghdad was the gateway for Muslim pilgrims to Makkah. Ibn Battuta's writing portrayed numerous buildings in Baghdad,

DOI: 10.4324/b23141-5

including schools, mosques, markets such as *suq al-thulāthā'* (Tuesday market) and the bridge of Baghdad. In 740/1340, hamd Allah al-Mustawfi visited Baghdad and wrote about it. He described the building of *al-madrasa al-mustansiriyya* as the most beautiful building in the city,[4] which points out the great devotion of people to learning venues. The intensification of travel literature in the 13th and 14th centuries shows some stability and safety in the region. The reading of these travelogues indicates the incentive to write was to express admiration for the unique attributes of Baghdad, and to contribute to general knowledge. Thus, the travellers' approach of this period can be considered as inspirational and appealing. The reputation of Baghdad's natural beauty and its high knowledge distinction continued to attract many explorers from adjacent cities. During the intermittent Ottoman rule of Baghdad, a number of Turkish explorers visited Baghdad and reported their remarks.

Among the eminent travellers of the 16th century is the Bosnian historian, teacher, geographer and cartographer, Nasuh Matrakçi, who visited Baghdad in 943/1537 with the Ottoman *sultan* Sulayman al-Qanuni.[5] He drew a unique map of Baghdad that shows both parts of the city, the river, the city wall, the bridge and a number of mosques and tombs. This map is an important document of 16th century Baghdad. It provides crucial information about the structure of the city and its components.[6] Al-Warid considers it the first complete map of Baghdad since its establishment.[7] This map highlights the river as if Nasuh imagined Baghdad to be a book lying open, with the river as the bookbinding (Figure 4.1).

In the 17th century, the Turkish historian Evliya Çelebi[8] visited Baghdad in 1066/1655. He wrote his observations about Baghdad's urban heritage in his famous book *seyahatname* (book of tourism), which consisted of ten parts. He observed there was a Sufi hospice occupying part of *al-madrasa al-mustansiriyya*, and noted Baghdad contained 6,000 wells and 100 waterspouts or *siqāyas*.[9] The *siqāya* is a water irrigation system that is usually established outside the mosques for people to drink for the love of God. These features were important parts of Baghdad's urban settings at that time. In 1066/1655 Çelebi visited Baghdad for the second time, and wrote an interesting description of its castles, markets and baths.[10] Also, another famous Turkish historian, Nazmi zadeh Murtaza visited Baghdad in the 17th century on his way to perform Hajj in 1100/1688. In the same year he finished writing his historical book *gulshan khulafa'*,[11] depicting the situation in Baghdad at the time of his visit, before the Mongol invasion. He described Baghdad as a paradise, that no other city bears a resemblance to its beauty, high castles, fertile soil and luxurious position.[12]

The large number of Turkish travellers designates the growing connection between Baghdad and the adjacent cities. People visited Baghdad as part of a

Figure 4.1 Sixteenth-century Baghdad by Nasuh al-Matrakçi [Jawad & Susa 1958].

trend of travelling, with a passion for further discovery of cities. In addition
to personal adventure, scholars travelled to exchange ideas and knowledge
with other scholars in the region. In the 17th and 18th centuries, explorers
from the neighbouring cities visited Baghdad as well, and documented their
observations. An example of those explorers is the Syrian scholar shaykh

Mustafa bin Kamal al-Din al-Siddiqi, who visited Baghdad in 1139/1726, and described many mosques, tombs and other unique buildings in his script.[13] Local travellers and historians continued to write their remarks, and they often utilised poetry to illustrate historical events. So, their texts contributed to history as well as to travel writing. Examples of these scholars include ʿabd al-Qadir al-Baghdadi (d.1093/1682), Ahmed bin ʿabd Allah al-Gurabi (d.1102/1690), ʿabd Allah al-Suwaidi (d.1174/1760), ʿabd Allah al-Fakhri (d. 1188/1774) and ʿabd al-Rahman al-Suwaidi (d.1200/1785).[14]

The incentive behind these writings was a desire to document different events to preserve their memory, in addition to contributing to general knowledge and appreciating the city's unique characteristics. So, this approach to travel writing can be outlined as an appreciative documentary approach. The large number of explorers who visited Baghdad throughout its history and documented their observations highlights the distinctive character that attracted such a huge number of visitors. It also shows the great attention to travel writing as an effective tool to learn about cities and their history. These travelogues took various shapes, including travel narratives, rhetoric, poetry, history books, city maps and architectural descriptions. Because of the similarity of social and political circumstances between Baghdad and other neighbouring cities, the explorers who visited Baghdad from the surrounding area experienced fewer difficulties in understanding the life and history of Baghdad compared with European travellers. Nevertheless, some of these writings were subject to political influences, and thus focussed on certain aspects and exaggerated real events.

European Travellers

It seems that only a few European travellers visited Baghdad before the 17th century, but the number greatly increased in the 18th and 19th centuries due to growing colonial interests. European travel writings reflected contrasting approaches: on the one hand they showed a curiosity for discovery and appreciation of place, yet on the other hand, they embodied an unfamiliarity and intolerance of differences. These attitudes influenced their conception of the city and was reflected in their writings. Italians were among the early European travellers who visited Baghdad. An example of these travellers is the Italian merchant, Cesar Federico who visited Baghdad in 1563.[15] Federico wrote his observations about Baghdad, and described the bridge, the houses and the alleys. Other Europeans, such as the French, Portuguese and Danish, were also interested in exploring 17th-century Baghdad. Among the emissaries of the 17th century was the Portuguese traveller Pedro Teixeira, who visited it in 1604 and wrote detailed remarks, reflecting his colonial curiosity.

Teixeira appreciated Baghdad's climate, noting "Bagdad enjoys a very pure temperature, and healthy climate".[16] He describes the bridge of Baghdad: "there is one bridge of twenty-eight boats overlaid with timbers, and between boat and boat is as much as the beam of one of them, that is, four paces".[17] He refers to the water of *dijla* (Tigris River) as "much clearer and sweeter than that of the Euphrates".[18] In addition, he noticed the city wall was "more than a league and a half about, and the other end rests on the river … there is a deep ditch all round, and the wall is of burnt brick, with platform, returns, and many bastions".[19] He observes "there is produced in the environs much cotton and silk; all wrought up and used in the city, where are more than four thousand weavers of wool, flax, cotton and silk, who are never out of work".[20] These products were all manufactured and used in Baghdad.[21]

In addition to the urban settings, Teixeira provided interesting remarks about the coffee houses of Baghdad, and how these gathering facilities were considered major places of entertainment for Baghdadis.[22] He comments "these places are chiefly frequented at night in summer, and by day in winter".[23] He also portrays the house that he resided in during his stay in Baghdad, which was overlooking the river, as "a very pleasant resort".[24] In general, he depicts the houses of Baghdad as "Large and roomy, but poorly built, and seldom well planned. All are flat-roofed; most have no windows on the street, and but small street doors".[25] He notes that almost every building in Baghdad was built with bricks, and that the use of stone is rare: "I do not remember having seen stone in any building of this city, except in the gateways of this *khan*".[26]

Interestingly, Teixeira illustrated the ruins of the round city of Baghdad, stating "for five miles around are found ruins of its great and fine buildings".[27] In addition, he indicated there were "visible in Bagdad ruins of fine buildings of the Persian times".[28] Similar to all other emissaries, he described the defence force, suggesting "the force appointed to the defence of this city … is commonly of fourteen thousand men",[29] Teixeira's broker was "a Jew turned Turk, in whom the Portuguese and Venetian men of business put much faith",[30] Teixeira's observations suggest a pleasing and healthy atmosphere in Baghdad. However, underlining political matters, more than other aspects of the city, reflected his preoccupied colonial intentions. He puts special emphasis on the defence force of Baghdad, specifying it in detail. Along with other European emissaries, he relied on a Jew to guide him while staying in Baghdad. This matter shows the great degree of unfamiliarity felt while there, and it indicates that many European travellers were either Jewish or they had alliance with Jewish firms.

Teixeira denotes the Muslim people of Baghdad as Moors and refers to the time of Islamic rule as a Moorish epoch.[31] The term 'Moors' primarily refers to the Muslim inhabitants of the Maghreb, Iberian Peninsula, Sicily and Malta during the Middle Ages. Although the Moors were not distinct or self-defined people, the name was later also applied to Arabs. To justify it, *Encyclopaedia Britannica* observed that "the term 'Moors' has no real ethnological value".[32] This term is one of the many tricky terms and labels implemented in European travelogues. Another emissary in the 17th century was the Italian traveller Deulofeu, who visited Baghdad in 1024/1616, and portrayed a pleasing image of its urban settings, especially markets that were full of silk. He ultimately enjoyed being in Baghdad and married a Baghdadi girl who joined him in his travel to Persia.[33] In 1042/1632 a French emissary, Tavernier passed through Baghdad on his way to India, and also visited Baghdad on his way back in 1063/1652. He wrote a comprehensive travel diary and provided a remarkable description of Baghdad.[34] He drew a map of Baghdad showing the city wall, the military castle and other buildings. In 1091/1680 a Danish emissary, O'Dyer, visited Baghdad and illustrated a map of the city in his travel book.[35]

This brief illustration of the history of travel writings before the 18th century shows various attitudes to travel writing throughout several stages in Baghdad's history. For regional travellers (explorers), the main objective of visiting Baghdad was exploring this great city and learning about it. Yet for European travellers (emissaries), their visit was part of their mission to inform their agents about the characteristics of the lands they visited and their capacity for future invasion. The material structure of Baghdad at that time wasn't described as amazing, compared with the round city of the Abbasids. Though Baghdad appears in the travelogues of this period as brilliant, prosperous and attractive, which escalates the need for further investigation of these texts, to discover more historical facts of this period.

Notes

1 Samarra' is a city in Iraq standing on the east bank of the Tigris River, 125 kilometres north of Baghdad.
2 For details see al-Attar, *Baghdad: an urban history through the lens of literature*, pp. 7–14.
3 For more details see al-Warid, *ḥawādith Baghdad fi 12 qarn*, pp. 122, 134, 143 and 151.
4 Al-Warid, *ḥawādith Baghdad fi 12 qarn*, p. 153.
5 The sultan visited Baghdad shortly after the Ottomans controlled it in 940/1534.
6 This map is taken from: Ayduz, S 2008, 'Nasuh Al-Matraki: a noteworthy Ottoman artist-mathematician of the 16th century', Viewed 5 January 2014, <MuslimHeritage.com>.
7 Al-Warid, *ḥawādith Baghdad fi 12 qarn*, p. 184.
8 He was also called Muhammad Zilli.

9   Sinclair, WF 1967, *The travels of Pedro Teixeira, with his 'Kings of Harmuz'
     and extracts from his 'Kings of Persia'*, with further notes and an introduction
     by Donald Ferguson, Hakluyt Society, Kraus Reprint, Nendeln, Liechtenstein,
     p. 52.
10   Al-Warid, *hawādith Baghdad fi 12 qarn*, pp. 202–203.
11   Al-Warid, *hawādith Baghdad fi 12 qarn*, p. 210.
12   Khuja, KA 2006, 'muqtaṭafāt min kitāb gulshan khulafa' by Murtaḍa N, *Arabic
     Translators International*, viewed 4 August 2012, <http://www.atinternational
     .org/forums/showthread.php?t=7467>.
13   Al-Warid, *hawādith Baghdad fi 12 qarn*, p. 217.
14   For more details about those historians see al-Warid, *hawādith Baghdad fi 12
     qarn*.
15   Al-Warid, *hawādith Baghdad fi 12 qarn*, p. 188.
16   Sinclair, *The travels of Pedro Teixeira*, p. 67.
17   Sinclair, *The travels of Pedro Teixeira*, p. 61.
18   Sinclair, *The travels of Pedro Teixeira*, p. 61.
19   Sinclair, *The travels of Pedro Teixeira*, pp. 63–64.
20   Sinclair, *The travels of Pedro Teixeira*, p. 67.
21   Al-Warid, *hawādith Baghdad fi 12 qarn*, p. 197.
22   Sinclair, *The travels of Pedro Teixeira*, p. 62.
23   Sinclair, *The travels of Pedro Teixeira*, p. 62.
24   Sinclair, *The travels of Pedro Teixeira*, p. 63.
25   Sinclair, *The travels of Pedro Teixeira*, p. 65.
26   Sinclair, *The travels of Pedro Teixeira*, p. 61.
27   Sinclair, *The travels of Pedro Teixeira*, p. 69.
28   Sinclair, *The travels of Pedro Teixeira*, p. 64.
29   Sinclair, *The travels of Pedro Teixeira*, p. 64.
30   Sinclair, *The travels of Pedro Teixeira*, p. 33.
31   Sinclair, *The travels of Pedro Teixeira*, pp. 65, 69.
32   Encyclopaedia Britannica <https://en.wikisource.org> viewed 10 March 2020.
33   Al-Warid, *hawādith Baghdad fi 12 qarn*, p. 196.
34   See Tavernier, JB, Crooke, W & Ball, V 1977, *Travels in India*, 2nd edn, 2 vols,
     Oriental Books Reprint Corporation, New Delhi.
35   For more details refer to al-Warid, *hawādith Baghdad fi 12 qarn*, pp. 202–208.
     The names of the travellers in the book of al-Warid are written in Arabic, and I
     transliterated them to English.

5 European Involvement in the 18th Century

The 18th century witnessed an increase in the number of English emissaries to Baghdad, as well as French and some German emissaries, which reflects a growing interest by the British and French in Baghdad and shows a diminishing interest by Italians and Portuguese. The writing of each European traveller contributed to Baghdad's history in diverse ways. The first English emissary who visited Baghdad in the 18th century is called Carsten. He reached Baghdad in 1145/1732 and wrote his notes.[1] In 1179/1765 the German emissary Neibuhr arrived in Baghdad and left in 1180/1766. He wrote a detailed description of the city's features. His most remarkable work is the map of Baghdad, which is considered the sole detailed map of late 18th-century Baghdad. In the same period, the English emissary Jackson visited Baghdad in 1181/1767, and he also drew a map of the city.[2]

From 1780 until the early years of the 19th century, more French emissaries arrived in Baghdad. In 1194/1780 a French doctor called Mashie visited Baghdad and made a unique discovery. He found an old Babylonian stone from the 12th century BC with cuneiform writing that illustrates the name *Bagdado* on it.[3] This discovery confirmed the ancient age of Baghdad. In 1195/1781 another French emissary, called Pushan, visited Baghdad and described the defence system. He also highlighted successful trade activities in Baghdad. The French emissary, Olivier, visited Baghdad in 1205/1791 and described its wall, bridge and markets. In addition, he wrote about the social aspects of Baghdad.[4] In 1215/1800, a French councillor, Russo, who was a friend of Pushan, arrived in Baghdad and wrote his observations.[5] The intensification of French emissaries in this period points out the strong will of the French to invade Baghdad. However, this interest had diminished by the mid-19th century, when they left it to the British upon a political agreement between them. Due to the limited capacity of this book, the travelogues of only two emissaries are discussed in the next sections, namely Niebuhr and Olivier.

DOI: 10.4324/b23141-6

Mapping Efforts of Carsten Niebuhr

The German emissary Carsten Niebuhr visited Baghdad in the second half of the 18th century. Niebuhr was born in West Ludingworth in 1733.[6] In 1760, he was invited to join the Arabian expedition of Frederick V of Denmark to Arabia Felix. This journey started in 1761 and ended in 1767.[7] The group consisted of six experts in different areas, and Niebuhr was the surveyor and the geographer of the group. Their proclaimed aim was "to study the customs, language, geography, flora and fauna of Arabia, and also to collect and study copies of the Biblical manuscripts … in Sinai".[8] The group started the journey by visiting Egypt, then they travelled to Yemen, Hejaz, Oman and India. Because of difficult travel conditions, all five of Niebuhr's colleagues had died by the time he reached Bombay. On his way back to Copenhagen, he passed through the Persian Gulf, Shiraz and then Baghdad.[9] He arrived in Baghdad in January 1766,[10] and was continually taking notes and measurements. From Baghdad he went up north to Mosul then to Aleppo, and lastly Damascus where he stayed for a few days.[11]

Soon after returning to his homeland, he compiled the results of his travels in his book *ravels through Arabia and Other countries in the East*. This book depicts the natural history of Arabia, including climate, fauna, flora and geology, in addition to people's manners, religion, character and governments.[12] According to Vernoit, Niebuhr's travelogues "brought travel writing to a new level of sophistication, especially with regards to archaeological investigation".[13] Buckingham noted that since Niebuhr's visit to Baghdad "there has been no traveller … who has had any opportunity of examining the country between the Euphrates and the Tigris".[14] Niebuhr produced a map of Baghdad that shows the two sides of the city, the city wall and ditch, gates, castle, markets and residential blocks. The map contains detailed plotting of the streets and wells, but only a few buildings are shown on the plan. The buildings that are shown on the map are the *sarāy*,[15] the *qal'a* (castle), *al-madrasa al-mustansiriyya*, *jami' al-Khulafā'*, some *sufi* hospices and some tombs. These buildings were presented in numbers on the map[16]. The reason for fewer buildings being illustrated on the plan of Baghdad, compared with the Cairo plan, was because of "Niebuhr's longer stay in Cairo, and the visit to Baghdad was towards the end of his travels".[17] However, this map became an important resource in the historical research on Baghdad, for it provided information about that special period (Figure 5.1).

Hopkins suggests "Niebuhr's most successful plans are probably those of Cairo and Baghdad".[18] On this map, Niebuhr located the site of the round city of Baghdad. The task of locating that city site was difficult because the ruins of the round city were hard to identify, due to continuous flooding, and

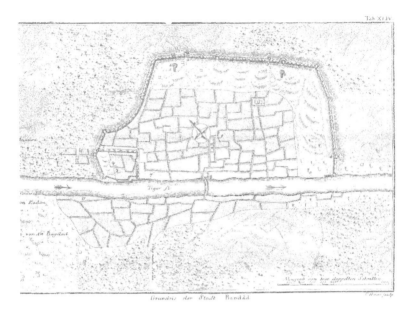

Figure 5.1 Eighteenth-century Baghdad by Carsten Niebuhr [Niebuhr 1983].

the accumulation of new buildings on the site. So, many travellers mistakenly suggested other locations for the round city. Similar to other emissaries, Niebuhr wrote about the surrounding area of Baghdad and highlighted some historical sites near it, such as dur-Kurigalzu from the 14th century BC, about 30 kilometres west of the centre of Baghdad, and the historical city of Mada'in that dates to 200 BC, located about 16 kilometres east of Baghdad.[19] He notes Baghdad had a large population compared with other cities in the East.[20] He indicates the majority of Baghdad's residents were Muslims,[21] with a few Christians and Jews.

Niebuhr highlighted the urban settings of Baghdad. He describes the city wall as weak, built of baked bricks. He notes there was a ditch outside the wall, yet it was dry, and there were no strong defence systems outside the city, enabling the Europeans to occupy it easily.[22] These notes reflect conquering attitudes in Niebuhr's writing. He portrays narrow roads inside the neighbourhoods of Baghdad. Conversely, the markets' pathways were wider, and covered with pointed arches. One of Niebuhr's concerns was the lack of bookshops, as he couldn't find a bookshop that sold books in Baghdad. He states the only way to sell old books in Baghdad was through auctions of deceased estates. He points out that Istanbul was the only place in the Ottoman Empire where Europeans could buy old books in the Arabic, Persian and Turkish languages.[23]

Regarding mosques and tombs, Niebuhr estimated about 20 mosques in Baghdad had minarets, and the rest were without. He was amazed by the decorations and calligraphy on the mosques and other public buildings, so he tried to copy some writings and include them in his book. The Arabic inscriptions were normally written above the mosque and tomb entrances (Figure 5.2). These writings usually embrace verses from the Qur'an, or a poem that documents the date of construction of that mosque, and the name of the governor at that time.[24] Niebuhr described the dome of the mosque of shaykh 'abd al-Qadir al-Gilani as great, but not magnificent.[25] In addition, he mentions other public buildings, such as a hospital, 22 khans and many *hammāms* or public baths.

Regarding the Tigris River, he estimated its width to be around 600–620 feet (183–189 metres). He designates the only bridge of Baghdad as weak, constituting 34 boats tied together with three strong chains. He refers to the topographic differences between the two parts of Baghdad, stating the eastern part was slightly more elevated than the western part.[26] This suggests elevated land was among the reasons for establishing the ancient market on that site. Also, it was a motive for choosing the eastern part to be the centre of subsequent governments over a long period in the history of Baghdad. The fertile lands between the Tigris and the Euphrates that

Figure 5.2 Arabic calligraphy on the main entrance of al-Sahrawardi tomb [Selman 1982].

were intersected by numerous canals amazed Niebuhr, who described them as "so rich a tract of country naturally invites its inhabitants to cultivate it".[27] He highlights the produce of the region of Baghdad, noting it was mostly wheat, rice and dates. For him, these kinds of food did not meet the 'Europeans desires',[28] maybe because they were not familiar with them. Further, he notices many Europeans who resided in Baghdad were medical practitioners who were also missionaries.[29] This indicates the political use of subjective needs, which has always been neglected in the pursuit of historical truths.

Moreover, Niebuhr elaborated on lofty courtyard houses that were built with bricks. Each house had a square courtyard in the middle, and most rooms overlooked the courtyard, because they had few or no openings to the road. He depicted the *sirdāb* (lower cooling room), and the *bādgir*, which is a device that allows air circulation inside the *sirdāb* to cool the room. It is extraordinary how this housing style, which originated in ancient Ur south Baghdad more than 3,000 years ago,[30] was maintained until the mid-20th century. This shows the long existence of courtyard houses as part of the area's architectural history and typology (Figure 5.3). Niebuhr's depiction of these houses as 'lofty' suggests an improvement in housing conditions with an increase in population, compared with single-storey houses of the

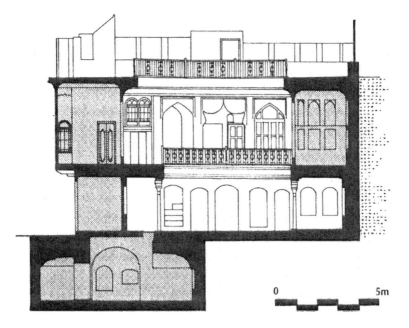

Figure 5.3 Unique housing typology of Baghdad [Warren & Fethi 1982].

early 18th century. Unfortunately, the unique style of courtyard houses has diminished greatly with the approach of 'modern' architypes that never consider specific requirements of place and its long architectural history. It is apparent that Niebuhr's remarks are only technical, and they are empty of emotions, unlike the writings of the explorers that are full of expressions and emotions.

Apart from the urban settings, Niebuhr was interested in the political affairs of Baghdad. He dedicated a huge part of his book to the history of governors, including Hassan *pāshā* and Ahmed *pāshā*, and he narrated other political accounts from the early 18th century. He recognises the establishment of the central committee that was responsible for discussing important issues and making decisions. The place where those high officials and religious scholars met was called *diwān Baghdad*. He points out this office was instituted during Sulayman 'abu Layla's rule (1749–1762),[31] and decisions made by this institute played a major role in appointing governors, collaborating with them and supporting them. Further, Niebuhr focussed on tribal issues, which are often connected to political issues. Historically, tribal Arabs are well known for their bravery, hospitality and generosity, and they would strongly defend their lands and their tribes. Niebuhr was anxious about his safety, so he dealt with rural people from outside Baghdad with superiority, calling them Bedouins: "all travellers complain of the robberies of these Bedouins".[32] Undoubtedly, travelling was dangerous everywhere in the world. Also, the issue of safe travelling is complicated because it is interrelated to specific political, economic and social issues of that era. Yet instead of relating an incident that he might have experienced, he generalises the problem. His irresponsible comments on people created a platform for differentiation that brought up many negative results which have lasted until the present day.

It is apparent that Niebuhr's concerns about European authority in Baghdad, their trade and their safety, influenced his travelogues. These concerns created a barrier to a pure travel experience that involves observing, finding and interpreting to seek additional knowledge about places. It seems that Niebuhr wanted his readers to obtain a remote sense of excitement from his adventures, by centring his writing on danger. Although it is obvious that he had found much beauty in his trip to Baghdad, it is often difficult to disentangle fact from fiction in his book. His personal observations about architectural qualities of Baghdad refer to the architectural style that organised the central concept of the system, rather than an architectural typology that established particular connection to place. Historians consider Niebuhr's work interesting "both from a historical and a scientific point of view".[33] However, the reading of his historical notes about Baghdad shows some confusing narratives that were transmitted vaguely without careful consideration of truthful references.

Examples of these accounts are the story of shaykh ʿabd al-Qadir al-Gilani and his pair of shoes, and the story of Bahlul and the two men.[34] Hence, his writings would have been more accurate if they had focused more on geography and topography instead of history.

The Novels of Guillaume Antoine Olivier

Olivier was a French emissary who travelled in late 18th century and wrote long novels about the places he visited. His visit to Baghdad was part of his obligation to the French government, to tour the cities that were under Ottoman control and write reports about them. He was born in 1756, and in his thirties he became famous as an entomologist and a traveller. He arrived at Baghdad in 1791, around three decades after Niebuhr's visit, and wrote his personal observations about this city. His travelogue pays considerable attention to the history of Baghdad, illustrating it from the time of its foundation. His two travel books: *Atlas to illustrate the travels in the Ottoman Empire; Egypt; and Persia*, and *Entomology or natural history of insects (1808)* and *the trip to the Ottoman Empire, Egypt and Persia*, were published in 1802, and 1808.

Olivier notes Baghdad had no parallel among the cities that were under the Ottoman control. He emphasises the great position of Baghdad in history, stating it was shining for five centuries as a capital of a huge empire, and a large centre for trade between the East and the West.[35] He implies even in the period following the Mogul invasion, Baghdad maintained its glory and remained an important trade hub and a central land port for Hajj caravans.[36] He also points out various political powers that took control of Baghdad throughout different periods in history. He refers to Baghdad as a Turkish city, yet in 1791 Baghdad was in the hands of the Mamluks who were partially independent from the Ottoman government until 1831.[37]

Although Olivier's remarks show appreciation of Baghdad's characteristics in history, his intensive focus on the historical accounts, and his superficial illustration of facts resulted in mixed impressions. His novels sometimes pay attention to specific aspects of the city's history, yet they overlook other important aspects, which resulted in partial meaning. Examples of this are the emphasis on the prosperity of the Abbasid period, and the superficial attention to political issues. Similar to Niebuhr, Olivier depicted the natural surroundings of Baghdad. He refers to the eastern part of Baghdad as 'Baghdad', stating Baghdad lies on a plain on the right bank of the river.[38] There is no doubt that the eastern part of Baghdad was more significant than the western part (*al-karkh*) as it comprised the administrative institutions. Though for the residents, both parts belonged to Baghdad. Narrations composed during the same period support this vision.

In addition to the surroundings of Baghdad, Olivier describes the city wall, noting Baghdad is bounded by a big deep trench and high walls that are well maintained. The wall contained many towers of different sizes, and had a wide base, becoming narrower at the top. He notes the city wall did not continue on the river's edge, because the houses were built on the river's edge, which was unusual, compared to other cities in the Ottoman Empire.[39] He also states Baghdad had four gates on the land sides and one gate from the river's side. The eastern part of the city used to have a defence wall on the river's edge for centuries. In the late 17th century, Baghdad's governor ordered the demolition of that part and to open the city to the river. The western part of Baghdad is identified by Olivier as 'western suburb', describing it as a heavily populated suburb that ended with the ruins of old Baghdad, or the round city.[40] According to him, the western suburb was also fortified with a wall that had defence towers and a trench, yet it was less complex than the eastern wall. The western part's wall does not exist in Niebuhr's plan produced in 1767. This suggests the wall was built in the late 18th century and indicates the increasing significance of the western part towards the end of the 18th century.

Regarding the demography of Baghdad, Olivier notes it was not heavily populated; it had large areas of unpopulated land in the eastern and southern parts of the city.[41] He estimated the population of Baghdad in 1791 at around 80,000[42] and he suggested the reason for the decrease in population was because of huge taxes imposed by *sultan* Murad in the mid-17th century, which drove people outside Baghdad. However, by the mid-18th century, more people in the region sought refuge in Baghdad because of terrible circumstances in their own cities, resulting in a subsequent increase in population.[43] The increase in population became obvious with the rise of two towns outside Baghdad. Olivier refers to *qaryat Musa al-Kazim* (village of *Musa al-Kazim*) on the western side to the north of the ruins of the round city, and to *qaryat al-Imam al-A'zam* which was on the eastern side of the river opposite the first town.[44] He implies the site of the round city was excavated by people to remove bricks and other building materials from the ruins.[45] The need for construction materials to build more quarters indicates the expansion of the city, and rapid population growth.

Baghdad's weather and its river's atmosphere impressed Olivier (Figure 5.4). He states the weather was very healthy, because Baghdad lies on a large plain that exposes it to constant wind, which helped to decrease the spread of diseases. He expresses the pleasure of staying in a house that overlooked the Tigris River. He also notes drinking water that came from *dijla* (Tigris River) was pure and healthy, and the sky was clear most of the time, which added extra beauty and comfort.[46] He recognised the extremely high heat during the day in summer is always followed by a pleasant breeze

Figure 5.4 The Tigris River atmosphere [Warren & Fethi 1982].

and moist air at night, saying fresh breeze increases the appetites and decreases the feelings of tiredness. He suggests these conditions are the reasons behind the magical effects of Baghdad's nights, which were always celebrated and enjoyed.[47]

Also, Olivier describes the boat bridge that connects the two parts of Baghdad. Comparing to Niebuhr who counted 34 boats along the bridge, he counted 30 boats instead. This shows a possible change in the size of these boats,[48] and different water levels that affect the stability of the bridge. Olivier suggested a strategy to sustain the beauty of Baghdad. He proposes if the lands surrounding Baghdad were all planted, and if the water of both Tigris and Euphrates rivers was well distributed to irrigate agricultural lands, there would be no place on earth healthier, livelier, richer and more productive, attractive and prosperous than Baghdad.[49] Unfortunately, due to continuous wars and irresponsible planning choices, present-day Baghdad has lost some of these beautiful features.

On the topic of architecture and urban forms, Olivier illustrated a fine picture of markets, describing them as the best components of the city. He notes the markets were wide and tidy, and they were enjoyable to pass through. They had high vaulted roofs to provide shelter, and at the same time deliver sufficient sunlight through small openings. In addition to markets, he described some significant buildings in both parts of Baghdad, including *al-madrasa al-mustansiriyya, jami' al-'aṣifiyya,* Talisman gate and the tomb of Zumurrud Khatun.[50] This tomb is a representative of outstanding construction techniques with brick and plaster (Figure 5.5).

While he enjoyed passing through wide markets, narrow and unpaved alleys in residential areas were unappealing for him.[51] Though, he portrays an efficient housing typology; the houses often consist of no more than two storeys, and the rooms are gathered on a perimeter of a square courtyard, which may contain some palm trees. He also points out important rooms, such as the basement room (*sirdāb*) and the big room on the second level, called *diwān*. This room was the main guest room, and it was used for the family as well.[52] It is usually a spacious room oriented to north and north-east directions to take advantage of favourable winds. These observations confirm the efficiency of such housing style, yet sadly they are not implemented in modern houses.

Moreover, Olivier wrote about the people of Baghdad, implying they were sweeter than others, and the elderly were more educated and pleasant. He observed great religious tolerance in the society, noting there was

Figure 5.5 The tomb of Zumurrud Khatun [Selman 1982].

no religious extremism in Baghdad, and envy was intolerable.[53] Further, he noticed that Baghdad's merchants were more effective in the society, and they were more devoted to the trading industry than other merchants in the Ottoman Empire. He commented on the level of honesty between the merchants to such a degree that shopkeepers could leave their shops open for some time without fearing burglars. These remarks indicate high values and stability in Baghdad society. Luckily this habit still exists between shopkeepers today.

It seems that Olivier's experience in Baghdad was pleasant. His observations painted a picture of a lovely environment that encompassed all measures of beauty, including social beauty. These comments reveal a significant perspective on the continuous prosperity of Baghdad after the Abbasid period, which is not highlighted in conventional historiography. Because of its unique location, the significance of its educational institutes, and the blessings of the river, Baghdad attracted many scholars, which may have added to those unique social qualities. The writing of Olivier and his detailed description of the urban forms of Baghdad and its natural environment is a good source for the material history of Baghdad in the late decades of the 18th century. However, similar to Niebuhr, his historical interpretations are often confusing, because they contain a mix of both truthful and misinterpreted meanings. Although invention is sometimes expected because of the nature of history writing, it is never accepted. Both Olivier and Niebuhr's writings on the history of Baghdad allowed more invention and contributed to its history representation problem.

Notes

1   Al-Warid, *ḥawādith Baghdad fī 12 qarn*, p. 219.
2   Al-Warid, *ḥawādith Baghdad fī 12 qarn*, p. 225. Al-Warid states this book has been translated to Arabic by Salim Taha al-Tiktrity.
3   Al-Warid, *ḥawādith Baghdad fī 12 qarn*, p. 228.
4   Al-Warrak 2007, *Baghdad biʾaqlām raḥḥāla* (Arabic), al-Warrak Publishing Ltd, London, p. 69. Also see al-Warid, *ḥawādith Baghdad fī 12 qarn*, p. 233.
5   Al-Warid, *ḥawādith Baghdad fī 12 qarn*, pp. 228–232.
6   Society for the Diffusion of Useful Knowledge 1833, *Lives of eminent persons*, Baldwin and Cradock, London, p. 2.
7   Vernoit, SJ 2007–2012, 'Niebuhr, Carsten', in *Oxford art online*, Oxford University Press, <http://www.oxfordartonline.com:80/subscriber/article/grove/art/T062405>, viewed 4 May 2013.
8   Scoville, S 1977, 'Beshreibung von Arabian by Carsten Niebuhr' (book review), *International Journal of Middle East Studies*, vol. 8, no. 2, pp. 275–276.
9   Hopkins, IWJ 1967, 'The maps of Carsten Niebuhr: 200 years after', *Cartographic Journal*, vol. 4, no. 2, pp. 115–118.

10 Al-Warrak, *Baghdad bi'aqlām raḥḥāla*, p. 20.

11 Niebuhr, C & Heron, R 1792, *Travels through Arabia and other countries in the East*, R. Morison and Son, Edinburgh, p. 178.

12 He uses the word 'Bedouins' to label Arabs in general as moving Arab tribes.

13 Vernoit, 'Niebuhr, Carsten'.

14 Buckingham, JS 1827, *Travels in Mesopotamia, including a journey from Aleppo, across the Euphrates to Orfah, (the Ur of the Chaldees) through the plains of the Turcomans, to Diarbeker, in Asia Minor; from thence to Mardin, on the borders of the Great Desert, and by the Tigris to Mousul and Bagdad; with researches on the ruins of Babylon, Nineveh, Arbela, Ctesiphon, and Seleucia*, Henry Colburn, London, p. xi.

15 The Ottoman administration office in Baghdad was called *sarāy.*

16 You can find this map in: Niebuhr, C (1733–1815) 1983, *Entdeckungen im Orient: Reise nach Arabien und anderen Ländern 1761–1767*, K. Thienemanns Verlag, Stuttgart.

17 Hopkins, 'The maps of Carsten Niebuhr', pp. 115–118.

18 Hopkins, 'The maps of Carsten Niebuhr: 200 years after', pp. 115–118.

19 Al-Warrak, *Baghdad bi'aqlām raḥḥāla*, p. 32.

20 Al-Warrak, *Baghdad bi'aqlām raḥḥāla*, p. 11.

21 Niebuhr relates Muslims to their prophet, calling them Mahomets, which indicates a superficial understanding of Islam.

22 Al-Warrak, *Baghdad bi'aqlām raḥḥāla*, p. 14.

23 Al-Warrak, *Baghdad bi'aqlām raḥḥāla*, p. 34.

24 Al-Attar, *Baghdad: an urban history through the lens of literature*, p. 77.

25 Al-Warrak, *Baghdad bi'aqlām raḥḥāla*, p. 19.

26 Al-Warrak, *Baghdad bi'aqlām raḥḥāla*, pp. 19–20.

27 Niebuhr & Heron, *Travels through Arabia and other countries in the East*, p. 173.

28 Al-Warrak, *Baghdad bi'aqlām raḥḥāla*, p. 34.

29 Al-Warrak, *Baghdad bi'aqlām raḥḥāla*, p. 36.

30 Warren & Fethi, *Traditional houses in Baghdad*, p. 18.

31 Al-Warrak, *Baghdad bi'aqlām raḥḥāla*, p. 48.

32 Niebuhr & Heron, *Travels through Arabia and other countries in the East*, p. 175.

33 Scoville, 'Beshreibung von Arabian by Carsten Niebuhr', pp. 275–276.

34 For more stories see al-warrak, *Baghdad bi'aqlām raḥḥāla*, pp. 19, 25.

35 Al-Warrak, *Baghdad bi'aqlām raḥḥāla*, p. 79.

36 Al-Warrak, *Baghdad bi'aqlām raḥḥāla*, pp. 82, 87.

37 Nawras, *hukk'm al-Mamālik 1750–1831.*

38 Al-Warrak, *Baghdad bi'aqlām raḥḥāla.*

39 Al-Warrak, *Baghdad bi'aqlām raḥḥāla*, p. 73.

40 Al-Warrak, *Baghdad bi'aqlām raḥḥāla*, p. 72.

41 Al-Warrak, *Baghdad bi'aqlām raḥḥāla*, pp. 73, 75. These remarks are similar to Niebuhr's notes, which shows different political and natural disasters that drove people outside the city.

42 Al-Warrak, *Baghdad bi'aqlām raḥḥāla*, p. 86.

43 Al-Warrak, *Baghdad bi'aqlām raḥḥāla*, p. 82.

44 Al-Warrak, *Baghdad bi'aqlām raḥḥāla*, pp. 83, 84.

45 Al-Warrak, *Baghdad bi'aqlām raḥḥāla*, p. 83.

46 Al-Warrak, *Baghdad bi'aqlām raḥḥāla*, p. 96.

47 Al-Warrak, *Baghdad bi'aqlām raḥḥāla*, p. 93.
48 Al-Suwaidi called these boats *sufun* (ships) which shows that they were big boats. See Ra'uf, IA (ed.) 1978, *tārīkh hawādith Baghdad wal-Basrah 1186–1192 AH, 1772–1778 AD* (Arabic), *by 'abd al-Rahmān al-Suwaidi*, The Ministry of Education and Arts, Baghdad, pp. 105–106.
49 Al-Warrak, *Baghdad bi'aqlām raḥḥāla*, p. 96.
50 Al-Warrak, *Baghdad bi'aqlām raḥḥāla*, p. 83.
51 Al-Warrak, *Baghdad bi'aqlām raḥḥāla*, p. 77.
52 Al-Warrak, *Baghdad bi'aqlām raḥḥāla*, pp. 75, 76.
53 Al-Warrak, *Baghdad bi'aqlām raḥḥāla*, p. 87.

6 British Intervention Intensifies

In the 19th century British colonial interests increased in Baghdad, with the establishment of the General English Consulate in Baghdad, in 1803.[1] Consequently, intensive tours were launched to Baghdad by English emissaries. A small number from other countries also visited Baghdad, such as the Russian emissary Tsyko Lela who visited Baghdad in 1223/1808 and wrote about it.[2] The English Resident in Baghdad, Claudius Rich, was among the early English agents who stayed there. Also, in 1231/1816 James Buckingham visited Baghdad and wrote notes in his book, *travels in Mesopotamia*, and in 1232/1817 William Heude visited Baghdad and wrote a detailed book about his journey.[3] In 1233/1818 an English painter called Kerr Porter visited Baghdad. In 1241/1825 an American emissary called William Fogg stayed in Baghdad as well, and pictured Baghdad as a city that "seems half buried in palm trees, which rise above the buildings in every direction".[4] This observation points out the significance of palm trees in Baghdad as a main urban component and as a vital urban emblem in its history. Other English emissaries include Robert Mignan who visited Baghdad in 1243/1827, and James Raymond Wellsted who visited in 1245/1829 and wrote the book *travels to the city of Caliphs*.[5]

When the Mamluks' rule in Baghdad came to an end in 1831 and the Ottomans regained their control of Baghdad, extensive British visits continued. Among the emissaries of this period was Fraser who visited Baghdad in 1250/1834, and Keith who stopped at Baghdad in 1253/1837, as well as Peril who visited it in 1257/1842.[6] In 1269/1853, a British official called Felix Jones surveyed Baghdad and produced a detailed map of the city.[7] In this map, he included 63 quarters in eastern Baghdad, and 28 quarters in the western part.[8] He affirmed the conditions of Baghdad had improved: "much of the city, which had been washed away by floods, had been rebuilt"[9] (Figure 6.1).

The list of the emissaries who went to Baghdad in the 19th century is a long one. In order to have a general understanding of the methods of

DOI: 10.4324/b23141-7

Figure 6.1 Baghdad in the mid-19th century [Susa 1952].

travel writing in this period, the travelogues of four emissaries are discussed in this chapter, namely Rich, his wife Mary, Buckingham and Heude. The official relations of Rich, the living atmosphere in Mary's letters, the descriptive manners of Buckingham and the colonial attitudes in Heude's writings constitute a variety of perspectives. Since all emissaries' minds were occupied with the image of Baghdad drawn by the *Arabian nights'* tales, it is essential to examine these tales before investigating travellers' texts.

The Arabian Nights

The image of Baghdad in the minds of European emissaries was stimulated by the tales of the *one thousand and one nights*, or what is known as *the Arabian nights.* This book consists of 1,000 stories that were written by anonymous authors in uncertain dates earlier in the ninth century. These stories were created for the purpose of entertainment for some kings in that period. The ninth century witnessed thriving aspirations for learning in the Islamic world, due to the wide spread of Islam, which allowed more mixing between many nations. Consequently, translation was increasingly encouraged to advance knowledge in many subjects. The increase in translation schemes resulted in a strong intermingling between different civilisations from the adjacent area around the Muslim world. Thus, some of the stories of *the Arabian nights* were influenced by Persian and Indian legends. Because Baghdad was then the centre of the Islamic world, many of those tales were inaccurately said to have happened in Baghdad. Although they comprise some facts, they are full of myths and imagination. Most importantly, these tales do not represent Islam and Muslims since they are strongly connected to the pre-Islamic era. However, European emissaries envisioned the Muslim world through these tales.

In 1704 a French antiquary and archaeologist, Antoine Galland, translated the *one thousand and one nights* and he became famous as the first European translator of these tales. His version of the tales appeared in 12 volumes between 1704 and 1717. Although he applied some changes and omissions to the original text, his book influenced the emissaries' perspectives of Baghdad at that time, and this was reflected in their writings. As a result, imaginative fiction occupied a large space in historical materials in regard to Baghdad and Islamic cities in general. For instance, when the English emissary Buckingham visited Baghdad in 1816, a century after the publication of Galland's version of the tales, he offered a large amount of money to buy an original copy of *the Arabian nights* but was disappointed when he realised it was difficult to obtain books and manuscripts. So, based on this single incident, he wrote that the literary scene was lifeless in Baghdad.[10] Equating the availability of one book to the intellectual status of Baghdad, while discounting the vast ground-breaking intellectual products of esteemed scholars, is problematic. It shows that the problem was in his limited capacity to understand the circumstances of bookselling in Baghdad, rather than a real shortage of books. In fact, scholars safely kept huge book collections inside their homes, and they would exchange or sell books whenever they needed to. In addition, the *Arabian nights* was not as appreciated by local scholars as it was appreciated by emissaries.

Although many emissaries expressed astonishment with the abstract imaginative image of Baghdad depicted in these tales, their real expressions were controversial. While some of them utilised the Abbasids' Baghdad to underestimate 19th-century Baghdad, others appreciated the urban settings of 19th-century Baghdad, considering the Abbasids' Baghdad as barbaric. The American author William Fogg, who visited Baghdad in 1872, asserts this attitude:

Having finished our tour of inspection, I said good-bye to my polite escort with a much better impression of the civilisation of Bagdad than I had ever before conceived of. Who will say there is no hope of future progress among a people where an hospital, an orphan asylum, and a printing office have been established, and are in successful operation!

And who would expect to find these evidences of refinement and civilisation in a city which is only associated in the minds of Europeans, as well as Americans with the barbaric splendour of the Caliphs in the time of the "*Arabian nights*".[11]

It looks that the emissaries were keen to find gaps in the account of Baghdad throughout history to serve their colonisation plans. Regardless of the contrasting images, the real denotation of Baghdad and its specific characteristics remained unidentified. Edward Said implies these traits created a strange variety of discourse a part of repetition and other representational issues of conventional historiography.[12] The dangerous aspect of these tales lies in their exertion of a huge negative influence on subsequent European literature and attitudes towards the Islamic world. This situation explains the danger of manipulation in history writing to serve personal goals, and the need to be aware of better techniques in interpreting specific texts, irrespective if they were interpreted in different ways.

Political Focus of Claudius Rich

The British Resident Claudius James Rich was among those emissaries who stayed in Baghdad in the early years of the 19th century. He arrived in Baghdad in January 1808, and he remained in this position as a Resident until 1821.[13] He was born in 1786[14] in Dijon in France, but moved to Bristol in England, as an infant.[15] In his teenage years he showed an interest in learning languages, and became "well-versed in Persian, Arabic, Syriac, Turkish and Hebrew".[16] In 1803, while he was only 17, he was appointed to the East India Company as a cadet. In 1808, he started his job as a British Resident in Baghdad.[17] By the end of his job in Baghdad he led excursions to the ancient sites of Babylon south of Baghdad and to the Kurdish region north-east of Baghdad. The narrative of this journey was published in 1836.[18]

It seems that Rich's colonial ambitions and attitudes, his personal contacts and his understanding of the Arabic language enabled him to be in that important position, even though he was young, only 22, and little educated. According to Bond, he was extremely arrogant;[19] he approached the people of Baghdad, including the governor, with superiority. Although the governance of Baghdad was in the hands of the Mamluk *pāshās*, he "was universally considered to be the most powerful man in Bagdad, next to the *pāshā*".[20] He married Mary, daughter of James Mackintosh, high court judge of Bombay. Bond observes "Rich had 'immunised' himself against the 'native' influence by marrying a traditional English lady".[21] So, even his marriage was part of his superior attitude.

Another example of Rich's arrogance is when he and his wife visited Basra in southern Iraq. On the day of their arrival, Basra's Resident, Manesty, who was married to a local woman, the daughter of an Armenian merchant, invited them to dinner. But Rich declined to bring his wife and wrote to Mackintosh; "I was much irritated by his presuming to mention Mrs Manesty, and to expect that I would permit Mrs Rich to associate with a dirty Armenian drab".[22] These awful words offer an insight into Rich's arrogance towards local people. This attitude complies with many Orientalists' approaches who "were often openly prejudiced".[23] Bond comments:

> One possible explanation for such attitudes is that the so-called 'Orientalists' considered themselves masters of the 'natives' as opposed to those who had 'gone native'. To 'go native' implied 'de-civilization' by the Orientals.[24]

Despite his disrespecting manners, Alexander presents him gently, recognising him as "one of the first to elevate the prestige of England in Mesopotamia".[25] Although many emissaries dealt with local people in the same manner, Rich had exceeded them all. He had "greater advantages than any other traveller",[26] because his official position enabled him to make authoritative decisions. Also, he stayed in Baghdad in a crucial time, with tensions between the Mamluk governor and the sultan, which allowed him to exploit these opportunities severely. With such personality, one can imagine the incentive of his actions that were celebrated conventionally.

Among the damaging approaches instigated by Rich as part of his attitude of supremacy were exaggerating the status of England and understating local people. As thought by Buckingham: "everything belonging to the Residency was calculated to impress ideas of great respect on the minds of the inhabitants".[27] This attitude enhanced discrimination and had psychological effects on people that diminished their confidence to control them easily. Simpson suggests making negative judgements is unusual: "in any age, it is not uncommon for people to rush to negative judgements about foreigners with different lifestyles, but 19th-century Europeans did so with a feeling of enormous assurance that they were correct".[28] So, the judgemental approach was implemented by other Europeans, including Rich. It seems that this approach was part of a psychological and imaginative tactic to contest their weakness. Adding to his corrupt behaviour, Rich illegitimately collected valuable antiques. Some were "Babylonian, and consisted of cylinders, amulets, idols, and intaglios, of the most curious kind".[29] Also, there were "gold and silver medals of the Sassanides, Sapor, and Ardeschair, collected at different periods, and many Cufic rings, seals, and talismans, with holy sentences engraved on them".[30] Extra collections included:

> A supposed seal of one of the Khalifs, dug up at Old Bagdad, and containing the words 'Ya Allah' O God! In large Kufic characters, deeply cut, on a substance resembling that of the ancient cylinders. A crystal seal, with Hebrew characters on it.[31]

The issue of antique collection raises a question about the lawfulness of the way they had been collected. Emissaries had no respect for the laws of ownership in Baghdad, nor to local residents. Quite the reverse, they used to complain about robbery on the way to Baghdad, ignoring the fact that collecting valuables illegally is robbery as well. Buckingham, who resided in Rich's house for a while, implies that Rich obtained some silver coins by digging up a container on the banks of the Tigris. He notes:

> [They] were obtained with difficulty by Mr Rich, as the *pāshā* wished to conceal the fact of treasure having been found in his dominions, from a fear that its amount would be exaggerated by the time the news reached Constantinople, and a demand of restitution from the Sultan might follow, as all treasures found in this way is his legal right.[32]

Although Rich was aware of the illegal aspect of antique collection, he proceeded in accumulating them, while sometimes pressuring the *pāshā* of Baghdad and beckoning him with money and gifts at other times. Similar to other emissaries, he was keen to obtain as many as he could of these antiques and artefacts, to transfer them to his homeland. This issue needs to be addressed to ensure justice and legal ownership of antiques worldwide. Nevertheless, it looks that his aggressive behaviour led him to losing the Resident job. In 1821 he was forced to remove the Residency from Baghdad after a dispute with the governor of Baghdad.[33] He went to Basra and then made a journey to Shiraz on his way to India, but he contracted cholera[34] and died at Shiraz shortly in 1821.[35] During his stay in Baghdad, Rich wrote several letters, which focused on politics rather than urban or geographic observations. These letters were not published during his short lifetime. Yet both his letters and his wife's letters were published later by Constance Alexander in her book *Baghdad in bygone days*. Based on Rich's letters, Alexander outlines the image of Baghdad:

> The city of the '*thousand and one nights*', its magical name conjures up a wonderful vision of Romance. Imagination paints the scene, a city of marble whiteness, from which rises the golden burnished domes

of many mosques ... or perhaps it is of palaces which hide the lovely beauties of a *pāshā*'s harem.[36]

Similar to other European emissaries, Rich and his wife Mary, approached Baghdad with an imaginative picture that is inspired by the book, *one thousand and one nights*, which goes back to the Abbasid period (168/785– 232/847). They connected that magic picture to their present time and envisioned the palace of the *pāshā* to be similar. Alexander points out the condition of Baghdad as Rich viewed it in 1808:

> The Baghdad of reality presented no spectacle. It was a dirty town with narrow streets and ill-kept walls. Its mosques and bazaars were many, but they had no outstanding feature or merit more than any other town of Turkey ... It was certainly a big trade mart and an important town ... The position of the town was the key to its suitability.[37]

The image suggested by Rich refers to the external material beauty of Baghdad in the early 19th century. Conversely, as stated before, the literature portrayed by Baghdadi scholars of the 19th century indicates the outstanding beauty of Baghdad. He highlighted the position of Baghdad, and its vibrant trade activities, among its positive aspects, yet he expected to see golden domes and spectacular architectural features. He did not realise the Abbasids' Baghdad had been subjected to political, social and natural factors that changed its appearance. In fact, the city of the golden domes was not ideal, because that supreme material beauty covered many social and political problems.[38] Rich's comparison between a superficial temporary image and an imaginative exaggerated image caused much misunderstanding and misinterpretation of the urban history of Baghdad.

The material characteristics of Baghdad in the early 19th century cannot be compared to the Baghdad of the Abbasids. Although some buildings were unappealing for Rich, many of them were very attractive from inside, which confirms that the spirit of the city and its internal beauty has not been changed. It is important to realise the interlocking criteria of cities in history in order to truthfully understand their conditions. The uniqueness of Baghdad at the time of Rich's stay presents other measures of beauty, including a beautiful environment, social and spiritual beauty, in addition to internal material beauty. While it is useful to observe the writings by Rich, it is risky to consider many of his assumptions as historical facts.

Lifestyle Conditions by Mary Rich

Rich's wife, Mary, was the eldest daughter of Sir James Mackintosh.[39] This man is described as a "plain and unadorned creature",[40] which indicates some compliance with Rich's behaviour. Mary wrote a number of letters while staying in Baghdad. Her letters contain significant comments about Baghdad in the early 19th century. In contrast to Rich's gloomy visions, Mary's letters revealed a better impression of their stay in Baghdad. She portrays their huge luxurious house in Baghdad where the family lived, which also contained the Residency offices, as:

> A large and handsome house perfectly in the Turkish style ... It consists of three different courts, one of which belongs to me ... and it is the most comfortable, retired part of the house. I have one large, handsome sitting room which we have made the library and breakfast parlour, and where I always sit and receive my great visits. There are no less than six other small, comfortable rooms, with a fine, large open gallery all around an open courtyard. These apartments are perfectly separate from other parts of the house, which I never visit till the evening, when business is over.[41]

Mary provided a fine description of the house they lived in. She points out a lovely atmosphere of the courtyard house, expressing her feelings of comfort and leisure in that house. The courtyard style was implemented in Baghdad for thousands of years.[42] The completely private internal space opens the house to the inside rather the outside. Also, the flexibility of design and possibility of full separation between its parts, designate great architectural sustainability that responds to people's requirements. Mary describes this style inaccurately as 'Turkish'. Perhaps she considered it Turkish because Baghdad was under the control of the Ottoman Empire. The resemblance in building styles between Baghdad and the adjacent cities is a result of their common history, their shared social circumstances, their weather similarities and the easy movement and connection between them, which enabled transmission of ideas. The courtyard house style originated in Ur south of Baghdad since the Sumerian period, more than 3,000 years ago.[43] Therefore, this style has primeval origins in the region of Baghdad. So, associating it with Turkish styles reflects a superficial perception that is based on momentary views rather than historical assessments.

It has been historically proven that the courtyard style is ideal housing in Baghdad. With the emergence of Islam, this style was maintained and developed further, since it preserves privacy, which is important for Muslims. According to many historical sources, this style reached its peak of beauty and significance in the 19th century (Figure 6.2). Yet, only a few emissaries highlighted it in the same manner as Mary, which suggests the perception of things depends on the experience of each individual. In contrast to Mary, Neibuhr did not enjoy the courtyard house, so he wrote that

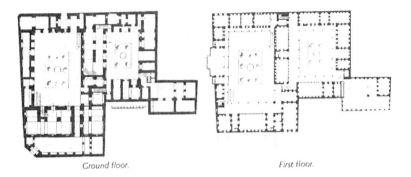

Ground floor.   First floor.

Figure 6.2 A plan of a big courtyard house in Baghdad [Warren & Fethi 1982].

the small square courtyard, which is surrounded by high walls, acts like an oven that increases the heat inside the house in summer. He adds that for this reason it was necessary for every building to have a lower underground room, or *sirdāb*.[44]

It seems that Neibuhr stayed in a small house in Baghdad during summer, and hot weather conditions affected his impression of these houses. Unlike Mary who lived in a huge house with big courtyards, the small courtyard house encouraged his negative thoughts. In fact, studies have proved that this style presents a brilliant architectural solution to solve diverse climatic conditions using available materials.[45] Apart from the luxurious house, Mary expresses delight in the quality and variety of fruit in Baghdad. She writes: "we enjoy a great deal of fruit; peaches, nectarine, apricots, apples, plums, and mulberries".[46] These statements show the excellent conditions of agriculture and trade in Baghdad at that time, as some fruit that are not usually grown in the Baghdad area were still sold there. However, Mary's reflection on the overall image of Baghdad was not content. She notes:

> The view I have of the renowned city is not the most beautiful. The streets are extremely narrow and the whole town is built of sun-baked bricks which gives it a very dirty appearance. There is nothing at all splendid in Baghdad, and the *pāshā* keeps up very little state.[47]

Mary's reflection of the external appearance of Baghdad was different to her delighted expression about the interior of their house and other buildings' interiors that sustained highly pleasing characteristics. Similar to Rich, the gloomy exterior picture for her contradicted the brilliant image outlined in the *Arabian Nights* tales. The change in Baghdad's conditions was a result of fluctuating political circumstances, in addition to natural causes throughout history. Also, it is narrated that the residents of Baghdad kept the exteriors unmaintained at that time, so they wouldn't appeal to the *pāshā* and his military who might confiscate good-looking houses.

Moreover, the narrow lanes between houses might have prevented enjoyable viewing.[48] Mary proposed another reason for the unappealing appearance; the governor of Baghdad was not keen to maintain the city structures. However, historical references verify a number of development projects in the same period, which suggests some exaggeration of the unpleasant physical condition of Baghdad. Regarding construction materials, Mary identifies the buildings in Baghdad as being constructed with sun-dried bricks. Similarly, another emissary, Heude, implies they were "unburnt bricks, or pieces of earth or clay".[49] Conversely, Buckingham outlines the bricks are baked and not sun-dried.[50] Mary and Heude could have referred to the old bricks that were reused, while Buckingham refers to the bricks that were manufactured in the 19th century. These contrasting views indicate different conceptions depending on the individual's knowledge and circumstances. However, comparative and collective analyses of these various texts would promote a better understanding of the historiography of Baghdad.

City Descriptions by James Silk Buckingham

Buckingham is among the English emissaries who stayed in Baghdad in the early decades of the 19th century. He arrived in Baghdad on 20 July 1816.[51] He incorporated his observations into three chapters of his book, *travels in Mesopotamia*. He was born in Falmouth, England, in 1787.[52] He received little education when he was young. He travelled to a number of cities in Europe, America and the East, and wrote books, as well as pamphlets on political and social issues. Before visiting Baghdad, he passed through other cities in the region, including Cairo, Aleppo, Damascus and Urfa. His perspective of Baghdad's architectural styles was placed in comparison to those cities. Sadly, this comparison sometimes ranked Baghdad behind,[53] because its physical image reflected its prolonged position as a frontier of political conflicts. He noted down his first impression of Baghdad:

> It seemed to stand on a perfectly level plain, it presented no other prominent objects than its domes and minarets, and these were neither so large nor so numerous, as I expected to have seen rising from the centre of this proud capital of the Khalifs.[54]

Buckingham implies the most dominant elements of the skyline were domes and minarets. Though these architectural elements weren't as large as he imagined, they shaped the city's identity. Along with all emissaries, he approached Baghdad with imaginative pictures from the *Arabian Nights* tales, which had an impact on his impressions, by limiting his understanding of the city spaces to a particular period and certain conditions. For example, he illustrates the *sarāy*, or the palace of the *pāshā* and his government offices, as "an extensive rather than a grand building".[55] And he notes the *sarāy* contains "most of the public offices, with spacious accommodations for the *pasha*'s suite, his stud, and attendants".[56] Before entering Baghdad, he experienced a problem with the customs at the gate of Baghdad, and he was detained for some hours.[57] When he arrived at Rich's house, he was very pleased with the lifestyle conditions, which he designates as "so much comfort, and, indeed luxury".[58] He stayed at Rich's house for a while. He describes the rooms and courtyards and states this residence "is certainly one of the largest, best and more commodious in the city".[59] This delightful experience delayed his task of visiting Baghdad. He wrote:

> I continued to enjoy these pleasures uninterruptedly for several days, before I felt even a desire to gratify that curiosity which is so generally impatient on entering a large and celebrated city.[60]

Similar to Mary, Buckingham highlights the unique characteristics of courtyard houses in Baghdad, especially if they are spacious, luxurious and comfortable. Staying at Rich's house was a good opportunity for him to explore these houses, as he had little access to other houses in Baghdad. He notes: "of the private houses of Baghdad I saw but little, excepting only their exterior walls and terraces".[61] This situation shows that he wasn't welcome in Baghdad, since he hasn't been to various houses. He portrays the houses' interior:

> The houses consist of ranges of apartments opening into a square interior court, and while subterranean rooms, called serdaubs,[62] are occupied during the day for the sake of shelter from the intense heat, the open terraces are used for the evening meal, and for sleeping on at night.[63]

These comments prove that courtyard houses are ideal, functional and they satisfy people's needs. Multiple apartment spaces embrace social and private activities with no clashes between them. Buckingham described the main entrance doors of these houses, indicating they had either round or flat arches, with fancy brickwork above them: "I had seen not even one pointed arch in the door of entrance to any private dwelling: they were all either round or flat, having a fancy-work of small bricks above them".[64] Although pointed arches were used in mosques, round arches were more popular in the houses, maybe because they were easier and cheaper to make. He illustrates that brick is the most-used construction material in Baghdad, because it can be manufactured locally, and it provides heat insulation. In addition, it is flexible and can be formed into fanciful and functional motives.

In regard to Baghdad's character, Buckingham states it "stands on a level plain, on the north-east bank of the Tigris, having one of its sides close to the water's edge".[65] Like Olivier, he refers to the eastern side of the city as Baghdad, though he mentions both sides. The eastern part of Baghdad drew the attention of travellers because it contained the government offices and was the largest part of Baghdad then.[66] Buckingham also noted that the western side of Baghdad had "same long continued line of streets and bazaars".[67] He also mentions a hospital on the western side, which indicates its remarkable growth during that period. Still, it was viewed as secondary compared to the eastern side. Buckingham points out a big part of eastern Baghdad, especially north-eastern areas, was not yet inhabited. The large uninhabited area inside Baghdad's walls is evidence of unpredictable population growth due to changing conditions. Nonetheless, historians suggest there was a constant increase in population in the second half of the 19th century.[68]

In relation to vegetation, he observed even populated parts by the river shores were full of trees, as if the city was built inside a forest.[69] The presence of plentiful vegetation indicates a great fertile land and a healthy environment. Buckingham also described the wall of Baghdad and its gates; "the wall is built entirely of brick, of different qualities, according to the age in which the work was done".[70] This wall "bears marks of having been constructed and repaired at many different periods".[71] Though the wall looked fragile to him, he states that Baghdad successfully resisted Persian attempts against it, and it was equally secure against the Wahhabis.[72] He observes that the old parts of the wall are more appealing than the new parts.[73] The decrease in the wall's architectural qualities is a result of recurring conflicts and natural disasters that forced people to quickly rebuild the damaged parts of the wall, without paying attention to fine details.

Further, Buckingham points out the city wall towers, some of which contained magnificent brick work, and long strips of Arabic calligraphy on top. He refers to Neibuhr's proposition that many calligraphy strips belong to the Abbasid period in 618/1221.[74] He notes there were three gates along the wall of Baghdad; one was on the south-eastern side of the city, and the second gate was on the north-eastern side. The third one, which is the main gate, was on the north-western side.[75] He did not indicate the fourth gate, *bāb al-tillism* or Talisman Gate, because it had been bricked up since the time of *sultan* Murad in 1638[76] (Figure 6.3).

Figure 6.3 Bab al-Tillism or Talisman Gate [Coke 1927].

However, he mentions a gate on the bridge that was called *bāb al-Jisr*, which implies this gate also existed then.[77] However, considering the north-western gate as the main gate indicates the city was accessible from the land further than from the river, which suggests a decrease in the use of ships as a means of transportation during that period.

The beautiful atmosphere of the Tigris River attracted Buckingham greatly. He notes the most impressive view that made him happy and relieved was the view of the two sides of Baghdad, as he stood in the middle of the boat bridge, where the dawn breeze was so quiet, and nothing could muddle its purity. The river flowed in grandness and splendour, and the sky with its glorious stars were reflected on the water's surface.[78] This image is comparable to the lovely images that are expressed in the literature and poetry by Baghdadi scholars, which indicates positive influence of the river's environment on both residents and visitors. Buckingham is among the few emissaries who transmitted this lovely image. Estimating the bridge of boats to be "little or more than six hundred feet in length", he was surprised that although the bridge seemed to be weak, the ropes and chains held it so well together.[79] Another appealing image of Baghdad for Buckingham was the pleasant social atmosphere during the nights of Ramadan – the month of fasting in the Islamic calendar – when main social activities usually take place at night. He narrates during one of these nights, while he was on the bridge, he could hear cheerful sounds from both sides of the river, which were well lit as far as the eye could see.[80] He notes:

> The large coffee house near the *Medrassee el Mostanser*, or College of the Learned, so often mentioned in the Arabian story, presented one blaze of light on the eastern side. The still larger one, opposite to this, illuminated by its lamps the whole western bank; and as these edifices were both facing the separate extremities of the bridge of boats, a stream of light extended from each, completely across it, even to the centre of the stream.[81]

The fascinating picture of the intertwining arrays of lights from both sides of the river and their beautiful reflection on the river's surface, reveals the ultimate beauty of the wonderful river's environment, and the two parts of Baghdad. The material beauty of this environment was undoubtedly impressive, yet it was the interconnection between this beauty and social and spiritual beauties of Ramadan evenings that offered such complete beauty and joyful atmosphere. These observations indicate a happier atmosphere in the early 19th century, compared with the late 18th century, and suggests a decline in epidemics and conflicts. Another lovely picture of Baghdad drawn by Buckingham was that of the city at midnight and early morning hours:

> But the scene that pleased me more than all, was that presented at midnight, from the centre of the bridge of boats across the Tigris. The morning breeze had, by this time, so completely subsided, that not a breath was stirring, and the river flowed majestically along, its glassy surface broken only by the ripple of the boats' stems, which divided the current as it passed their line.[82]

This remark depicts an amazing picture of the natural environment of Baghdad, which collectively encompasses life on the river's edge, people, boats, breeze and pure water. Unlike the pleasing effects of the environment of Baghdad, the city interior wasn't appealing for him, because it contained less interesting objects.[83] This also reflects the influence of *he Arabian Nights* on his imagination that painted a picture of unparalleled grandeur and splendour. In relation to building materials, Buckingham notes:

> All the buildings, both public and private, are constructed of furnace-burnt bricks, of a yellowish red colour, a small size, and with such rounded angles as prove most of them to have been used repeatedly before, being taken, perhaps, from the ruins of one edifice to construct

a second, and again, from the fallen fragments of that to compose a third.[84]

Buckingham implies that bricks were popular materials for construction. They were small and had rounded edges instead of sharp angles, which indicates their recurrent use. He designates the use of new bricks was uncommon, and if used they would be noticeable.[85] The practice of constructing buildings with used bricks indicates the existence of some ruins due to various events in history. Used bricks were undoubtedly cheaper and available, because of persistent floods that caused the demolition of some buildings. Recycling bricks sustainably saves efforts and makes the city cleaner, and this is among the good lessons of this experience. In another comment, he portrays the streets:

> The streets of Baghdad, as in all other eastern towns, are narrow and unpaved, and their sides present generally two blank walls, windows being rarely seen opening on the public thoroughfare, while the doors of entrance leading to the dwellings from thence are small and mean. These streets are more intricate and winding than in any of the great towns of Turkey, and, with the exception of some tolerably regular lines of bazars, and a few open squares, the interior of Bagdad is a labyrinth of alleys and passages.[86]

Buckingham notices the streets were narrow and unpaved, lined with high walls that had no openings to the ground floor level. He also notes the street network was complex, because the lanes were more winding than any other big city in the Ottoman Empire. The tightness in Baghdad's streets helped to cool the area and it reduced excessive heat in summer and ventilated the air. Also, extra twists in these lanes increased visual delight and reduced monotonous views.[87] So, having narrow lanes was beneficial and favourable for the residents of Baghdad (Figure 6.4). In contrast to the narrow lanes, Buckingham states that the markets were numerous, and mostly formed of long, straight and reasonably wide avenues. He predicts some of them belong to the 14th century, and they are well furnished with Indian commodities.[88] He thoroughly describes several features of markets, including roofing:

> The best of these are vaulted over with brick-work, but the greater numbers are merely covered by flat beams, laid across from side to side to support a roof of straw, dried leaves, or branches of trees and grass.[89]

This depiction highlights some physical components of markets. Vaulted brickwork markets were usually built in the central areas, and they were

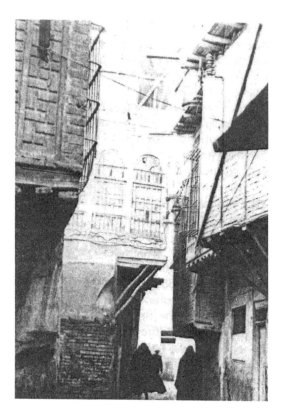

Figure 6.4 Narrow lanes between houses in Baghdad [Warren & Fethi 1982].

few, maybe because these roofs demand more time to finish, and it requires professionals to produce high quality brickwork. The use of palm tree branches for roofing is common in Baghdad, yet the use of grass is not as popular. In addition to roofing, Buckingham observed something special in these markets, noting it is "a peculiarity which I had never seen elsewhere".[90] The somehow different style in Baghdad denotes its specific historical situation as a collective place for many civilisations and backgrounds. He describes it as:

> A band of old Arabic inscriptions over each shop-bench, sculptured in large characters, and with as much care as any of the inscriptions on the mosque. These were executed with so much regularity and uniformity, as to induce a belief of their being coeval with the bazaar itself, which was very old.[91]

Traditionally, individual shops usually hold verses of the Qur'an, in order to obtain blessing and protection. This tradition shows people's adherence to the teachings of the Qur'an, with a desire to immerse its teachings in their daily activities. Unlike the inscriptions on mosques that incorporate both Qur'anic verses and poems, individual shops only had Qur'anic verses. This case proves the significance of inscriptions that are applied to monuments as well as ordinary spaces. In addition, it shows the permanency and long existence of markets (Figure 6.5). Buckingham couldn't recognise those scripts, so he asked a local person to explain them, which shows the importance of language to comprehend the meaning of place. Another condition of the markets of Baghdad instigated mixed feelings of surprise and pleasure inside him. He stayed in Baghdad through the holy month of Ramadan, in August, which is the warmest time of the year in Iraq. He was surprised to see little activity in the markets throughout the day, because people couldn't bear the excessive heat while fasting, yet these markets were very crowded and full of lights and life at night. Although he was not familiar with these conditions, he enjoyed the illuminated places at night:

> The bazaars, which were mostly deserted during the day, thronged at night by multitude of idlers, all arrayed in their best apparel … the peculiar gloom, which regained throughout these dark brick vaulted passages during the morning, was now removed by a profusion of lamps and torches, with which every shop, and bench, and coffee shed was illuminated and all was life and gaiety.[92]

Figure 6.5 A marketplace in Baghdad [Fogg 1985].

These remarks reflect dual images of the same urban component of the city. While the markets were lifeless during the day, they were full of joy and happiness during the night. This situation occurs only in Ramadan, especially if it happens to be in summer, when the temperature soars. Even though the market was materially beautiful, the lack of social beauty made it less pleasant, which indicates the importance of integrating social beauty with other measures of beauty. Buckingham inconsiderately describes people as 'idlers', which verifies his judgemental mind on people without trying to understand their situations. Unfortunately, such discriminatory statements created so much misperception in history. Nonetheless, his designation of the shift in social conditions of Baghdad during Ramadan is distinctive. Perhaps no emissary had experienced this situation and recorded it in their book. The sudden change of the atmosphere of the market during the day, before and during Ramadan, was unexpected situation for Buckingham.

This incident emphasises the law of continuous historical change, which is a normal occurrence that affects the interpretation of place. It also implies that judgements should always consider openness, to enclose all changeable elements of history, since temporality of everyday events resists the fixed images provided by conventional histories. A number of coffee shops within the markets were particularly portrayed by Buckingham. The large number of coffee shops indicates their increasing social value (Figure 6.6).

In addition to markets, Buckingham portrays the central open public space, or *al-maydān*, as an eventful place that "never failed to be crowded every night with people of all classes, and every mode of diversion in use here".[93] He highlights the happy atmosphere in this public square where people talk, discuss, and sing; "with blazing fires, lamps … to add to the effect of the general rejoicing".[94] Moreover, he recognises more than 50 baths in Baghdad, yet they were "inferior to those of all the large towns of *Mesopotamia*, through which I [he] had passed".[95] These observations demonstrate a beautiful city with regular social activities.

Buckingham estimates there were more than 100 mosques in Baghdad, yet he only estimates 30 mosques that could be distinguished by their particular minarets; "the rest are probably mere chapels, oratories, tombs and venerated places, restored by the populace for prayer".[96] The general mentality about mosques associates them with minarets. However, having a minaret is not always necessary; it depends on the size of the mosque and its location. People always need to pray in a nearby location; that's why mosques are always connected to the markets, regardless of whether or not they have minarets. Buckingham notes: "the mosques, which are always the prominent objects in Mohammedan cities, are here built in a different style from those seen in most other parts of Turkey".[97] Along with all emissaries, he calls Muslims Mohammedan, in relevance to the prophet Muhammed.

Figure 6.6 Outdoor coffee shop [Stark 1947].

It looks like this was the trend at that time, which reveals the changeable nature of some terms depending on the objective of their usage.

Moreover, Buckingham elaborated on the architectural styles and designs of some of the great mosques and coffee houses on each side of the Tigris River, such as *jami' suq al-ghazl, jami' al-khāsaki, jami' al-wazīr, jami' al-maydān* and *jami' Mirjān*.[98] He notices some of these mosques had "a fine appearance from without",[99] and he shows an appreciation of certain parts of these mosques, such as their entrances and minarets. For example, he depicts *jami' al-wazīr pic*, "which is seated near the Tigris, and only a few yards from *bāb el Jisr*, or gate of the bridge, has a fine dome and lofty minaret".[100] This comment affirms the gate of the bridge existed during Buckingham's visit to Baghdad, but it was removed later. The central mosque in *al-maydān* was also mentioned by Buckingham, portraying it as "a noble building".[101] This mosque is located on the way from the north-west gate to the palace and the British residence. Another mosque is *jami' Mirjān* that "has some remains of equally old and very rich Arabesque work".[102] Also *jami' al-khāsaki* had "a small portion of the original edifice remaining",[103] and it had "several Arabic inscriptions, in a good upright character, and one in the loose and flowing character of the Persians".[104] The main characteristics of these mosques are loftiness, centrality, durability and beautiful inscriptions. Although many of them were rebuilt in different periods, they maintained their qualities and spirit. These observations designate the old age of many mosques in Baghdad with different styles and motifs.

In general, Buckingham observes the domes to be high and narrow, with a height that is double the radius (Figure 6.7). He notes: "they are richly ornamented with glazed tiles and painting, the colours used being chiefly green and white".[105] The glitter of the colours of the domes "reflected from a polished surface, gives a gaiety and liveliness, rather than majesty or magnificence, to the building".[106] However, these domes appeared to him "much inferior to the rich and stately domes of Egypt, and especially those of the Mamlouk sepulchres at Cairo".[107] Unlike the huge number of symbolic references and multiple interpretations of mosques and domes expressed by Baghdadi scholars in the same period,[108] Buckingham's interpretation of mosques is limited to visual appearance, and it focused on their exteriors, which implies he could not gain access inside them.

In regard to the population of Baghdad, he estimated from 50,000 to 100,000, then he favoured 80,000, noting "maybe near the truth".[109] The population of Baghdad at the same time was estimated by another emissary, Heude, to be 200,000.[110] This huge difference confirms these numbers cannot be taken as historical statistics since they were only estimates. However, they can inform urban history studies of Baghdad. Furthermore, Buckingham brings out a description of different kinds of people and their religions. He was impressed with the local customs, and he specified the specific clothes of each group in Baghdad society.

Figure 6.7 High and narrow dome in a Jami' in Baghdad [Coke 1927].

Having explored Buckingham's writing in relation to Baghdad, we realise he had documented every observation in his diaries. Historians recognise him as a prolific author and "an observant and entertaining writer".[111] Nevertheless, because he had little access to private spaces, his writing on Baghdad focused more on the overall image of the city, and on public places. Also, since his task was merely to record what he sees, he wrote the comments without further investigation or analysis. Therefore, his writing could be considered useful and elaborative, yet they are partial and lacking visions into important aspects of Baghdad.

Colonial Interests in Heude's Writings

In 1232/1817 William Heude arrived in Baghdad as part of his journey from India to England. In this journey he visited Armenia, Asia Minor, Persia and Kurdistan, in addition to Baghdad. He collected his notes in a book called *a voyage up the Persian Gulf and a ourney overland from India to England*, which was published in 1819. He had connections with the Madras Military Establishment,[112] and was apparently related to Earl Fitzwilliam.[113] His stay in Baghdad was short, but he notes: "it had been marked by great events, and by one of those sudden revolutions".[114] By citing 'great events' he refers to a conflict between the Mamluks and the Ottomans over the sovereignty of Baghdad. He expressed some positive views about Baghdad, relating it to its great history: "it presents something in the novelties of the scene, but more in those recollections which are allied to the memory of its former condition".[115] He brings up some unique characteristics of this city:

> Its fruits as delicious and highly flavoured as any of the east; its people as highly civilised and perhaps more courteous in their demeanour towards Europeans than those of any other Mahomedan city we are acquainted with. It possesses many advantages both of climate and situation, and enjoys a considerable trade.[116]

These comments show a beautiful atmosphere, and they reveal exceptional characteristics of Baghdad, including its produce, people, climate, location and trade activities. In addition, the markets of Baghdad were very appealing to Heude; he states, "its bazars are splendid beyond anything we have ever seen in other parts".[117] Also, he commends the position of Baghdad in history:

> If we look back to the remotest ages, no country, I believe, will be found so much celebrated in the history of mankind as that now desert tract, which the Tigris and the Euphrates enclose within their banks.[118]

However, while admiring the features of Baghdad at the time of his visit, he suggests Baghdad's prestige is highly connected to its history: "still, if it were not Bagdad, the city of wonders and romance, it might not be so highly spoken of as it generally is".[119] Along with other emissaries, the reputation of Baghdad as a capital of the Islamic world influenced Heude's perception of this city. He considers the round city of Baghdad among those 'long lost cities' where the visitor would "naturally associate the recollection of their former state, with those faint, yet speaking traces of a past existence which they still exhibit".[120] He appreciates the survival of the round city traces at that time:

> Travellers enjoyed opportunities of observing the existence of those distinct traces of former grandeur, which sufficiently distinguish the respective situations of three great cities, now swept away by the desolating power of time. Babylon, Seleucia and Ctesiphon, have all gone by; Bagdad alone remains.[121]

Like Buckingham, Heude stayed in Baghdad at a time when Rich was the British Resident, which indicates an excessive number of British emissaries in early 19th century period. Although he wrote about Baghdad's natural environment and other urban elements, he focused extremely on the political status of Baghdad, describing it as 'a frontier town'.[122] For instance, he discoursed the situation of the Mamluks, noting a great number of *pāshās* who had governed Baghdad were "terminating their reign and their existence at once by a violent catastrophe ... lust of dominion and authority, that there are frequently more competitions than vacancies".[123] It is true the position of the governor of Baghdad was very attractive, yet the 'lust of dominion and authority' is a general attitude of all political powers in the world. It looks that he used these terms to criticise the Mamluks particularly, to establish negative impressions against them, to arrange for easy future control of the city.

In his book, Heude provided a detailed description of the people of Baghdad, stating the majority were Muslims, and of Christians of various sects he "could only learn that there were about one hundred and sixty Roman Catholic families",[124] in addition to some Jews. Like other European emissaries, he refers to Baghdadis as Bedouins. On one hand, he recognises their good ethics, describing them as 'our hospitable friends' with "their character, and manners, and observances that have been the leading characteristics of this celebrated race from the most ancient times".[125] Although these remarks seem positive, unfortunately, as suggested by Simpson, almost all emissaries relate good ethics to primitiveness, referring to "that sweet innocence of expression which is known only among a primitive people".[126]

On the other hand, he proposes they lived in 'half-barbarous climes'.[127] The term 'barbar' is used to denote specific tribes who lived in North Africa, called Amazigh. There are various suggestions to the root of this word; some suggest it has been derived from ancient Greek's thoughts "to describe all non-Greek-speaking peoples".[128] Others suggest it might refer to the Arabic word '*bar*' which means an outland, or it might relate to the Arabic verb '*barbara*' which means 'to say something that cannot be understood'[129]. This definition complies with the Greek's use of the word. A third opinion is that this term recounts some sounds of an angry person.[130] It is true that anyone who experiences oppression or invasion becomes angry, and therefore he could speak of words that are strange for foreigners. Hence, the origins of this term are different from the way it was used by the Europeans in the 18th and 19th centuries.

The Romans were the first to transform the meaning of this term. Yet, according to Fruitt, based on their definition of the word they "were barbarians themselves".[131] Later, when European emissaries wrote their travelogues, they linked barbarism to 'other' cultures, especially Turkish and Arabic, to designate them as primitive, and to intrude upon their civilisation and religion. Today, this term has been commonly used to refer to cruel acts from uncivilised people, which shows "how far 'barbarism' has been removed from its ancient roots",[132] by incorrectly shaping it with aggression. This case is among the many cases of misuse and diversion of language to serve political goals. Heude also used the word 'barbarous' to portray a conflict that he witnessed between the Mamluks and the Ottomans over the sovereignty of Baghdad: "for five successive days, this great city presented the same aspect of barbarous strife, confusion, and dismay".[133] It is expected to experience fierceness between military powers, yet the use of this term gave the impression of uncivilised atmosphere, which served his ambitions.

Another topic expanded by Heude is the Arab race. As stated by him, this race forms "one of the three distinct races that occupy Asia in general".[134] He designates an Arab man to be "quick in his emotions, sudden in his anger, restless, and bred to war, with a body of iron and a soul of fire … scarcely capable of the enjoyment of repose, except in those moments of satisfied impatience".[135] This stereotypical attitude is never realistic since people differ regardless of their race. Conversely, he conveys some moral characters of the Arabs noting they have "inherent weight of dignity",[136] and they are "extremely devout, and yet more tolerant and less assuming in their demeanour towards Christians".[137] A relevant issue that appeared in Heude's writing is associating Arabs with the desert. He implies: "the desert and freedom are their inheritance".[138] It is true there are some deserts in the area, yet Baghdad was full of vegetation and palm trees.[139] Associating

Arabs with the desert is an invention in history that had been implemented to accuse Arabs of backwardness.

Simpson explains the selective attitude of the European travellers, who were "often stuck to the stereotypes they carried in their baggage; it was natural for them to single out for emphasis the sights and scenes that confirmed their beliefs, and to overlook those that contradicted them".[140] Many of their fabricated claims "were shaped long ago by political, military and commercial conflicts that go back even beyond the Crusades".[141] So, the Arab race issue ascended from a long period of distrust, a few centuries before the 19th century. Although these thoughts are dangerous because they enhance extremism, they were normalised by repetition and by other political tricks. On another occasion, while Heude agrees about the cleanliness of the Arabs, he criticises other emissaries who greatly commended their cleanliness, because that would give credit to the Arabs:

> The personal cleanliness of the Arabs, the Turks, and other Asiatics, has often been spoken of, but we are by no means inclined to accord with what has been said of them by other travellers, who, in the profusion of their commendation of the habits and manners of these foreign climes, are sometimes apt to undervalue our own customs and observances, and to give them credit for those good qualities they certainly do not possess. In my own opinion ... the Arabs may not be quite so filthy as some European nations have been esteemed; but most certainly, in regard to personal cleanliness, they are not to be compared with ourselves; their frequent ablutions, where water is sufficiently abundant, very imperfectly removing the evils attendant on the length of time they wear their clothes.[142]

These statements show how Heude was intentionally not interested in praising the Arabs, whom he calls 'Bedooins',[143] fearing this would make them proud. These attitudes resulted in a mix of invented ideas to serve a specific goal, often occupation. He confirms his intentions:

> I am aware the contrary of this has been asserted in the praise which has been bestowed upon them, and which would allow of no drawbacks on the perfection of their character, and the superiority of their manners. I speak however, from my own observation.[144]

It seems that Heude looked forward to the break-up of the Ottoman Empire, while promoting the establishment of European colonies. He puts forward a whole plan for the British army to invade Baghdad, based on the information that he provided:

With these advantages a small army might easily reach Bagdad from Korna within the month; whilst our artillery, in the same time, might proceed up the stream in boats of a sufficient size to be formed occasionally as a battery for the reduction of the place ... with two or three of our brigades, to attack a city containing a population of near two hundred thousand souls.[145]

He also planned how to reduce the supplies of the army in Baghdad:

Thirty thousand men from Bagdad could not keep together for a month in the desert, through the want of supplies, unless they could command the navigation of the rivers; as advantage.[146]

Although he proposed the invasion plan around 1819, the real British invasion happened about 100 years later. This case shows the great danger of proposing destructive ideas and the long span of them turning into actions. In addition to the invasion plan, Heude discussed the matter of resistance by the people of Baghdad, implying that until they reach Baghdad, the "troops need not have any serious resistance to apprehend".[147] However, he expected the primary resistance would be by "an infuriated populace making common cause with their government against the enemies of their faith".[148] He expected people to be enraged, and he was afraid that they would be on good terms with their government against British forces. So, he suggested strengthening British relations with different tribes to create a 'constant dread'[149] of seeing them openly supported by the tribes. He willingly admits their forces are considered enemies by the people of Baghdad, and that the British may open their guns against people, which paints a horrible inhumane scene.

On top of all that, he endorsed sectarianism by consolidating religious differences that had never run greatly in Baghdad society. He advocates "they have certainly left that degree of reciprocal mistrust of which an enemy might always avail himself".[150] These comments indicate his will to find some gaps in Baghdad society, to expand them to serve his goals. He expected the castle could be gained "within the first or second day of the opening of our [their] guns",[151] which suggests he had no admiration for peoples' humanity, safety and wellbeing. While Heude's invasion plan reflected his keenness to make them successful, they indicate his fear from people and their capability to defend their lands. These psychological concerns normally occur with any kind of illegal intrusion. The plan clearly reveals his real intention of visiting Baghdad, to arrange for an occupation rather than a mere exploration of this city. Indeed, he had excelled all other emissaries by insistently disclosing his immoral intentions through his writing. Such intentions distorted the real meaning of travel and muddled up the response to the question of how pure travellers' attitudes and experience should be.

Notes

1 Al-Warid, *ḥawādith Baghdad fi 12 qarn*, p. 233.
2 Al-Warid, *ḥawādith Baghdad fi 12 qarn*, p. 234.
3 See Heude, W 1970, *A voyage up the Persian Gulf and a journey overland from India to England in 1817*, containing Notices of Arabia Felix, Arabia deserta, Persia ... Bagdad, Koordistan, Armenia, Asia Minor, London, Knightsbridge.
4 Fogg, WP 1985, *Arabistan, or the land of the Arabian nights: being travels through Egypt, Arabia, and Persia, to Bagdad*, Darf, London, p. 232.
5 See Wellsted, JR & Ormsby L 1968, *Travels to the city of the Caliphs, along the shores of the Persian Gulf and the Mediterranean: including a voyage to the coast of Arabia, and a tour on the Island of Socotra*, 2 vols, Gregg International, Farnborough, Hants.
6 Al-Warid, *ḥawādith Baghdad fi 12 qarn*, pp. 238–244.
7 Jones, JF 1998, *Memoirs of Baghdad, Kurdistan and Turkish Arabia*, 1857, selections from the records of the Bombay Government, no. xliii, new series, Archive Editions, Great Britain. Also see Jawad & Susa, *dalil khāriṭat Baghdad al-mufaṣṣal*.
8 Jones, *Memoirs of Baghdad, Kurdistan and Turkish Arabia*.
9 Dabrowska, K & Hann, G 2008, *Iraq then and now: a guide to the country and its people*, Bradt Travel Guides, Chalfont St. Peter, p. 199.
10 Al-Warrak, *Baghdad bi'aqlām raḥḥāla*, p. 145.
11 Fogg, *Arabistan, or the land of the Arabian nights*, pp. 312–313.
12 Said, *Beginnings*, p. 302.
13 Rich, CJ 1836, *Narrative of a residence in Koordistan, and on the site of ancient Nineveh: with journal of a voyage down the Tigris to Bagdad and an account of a visit to Shirauz and Persepolis*, 2 vols., J. Duncan, London, p. xxv.
14 Alexander, CM 1928, *Baghdad in bygone days: from the journals and correspondence of Claudius Rich, traveller, artist, linguist, antiquary and British resident at Baghdad, 1808–1821*, 1st edn, J. Murray, London, p. 1. In another source the author suggests that Rich was born in 1787; see Rich, *Narrative of a residence in Koordistan*, p. xv.
15 Alexander, *Baghdad in bygone days*, p. 1.
16 Bond, XW 2016, 'Claudius Rich and Samuel Manesty' <https://blogs.bl.uk/untoldlives/2016/03/claudius-rich-and-samuel-manesty.html> viewed 3 September 2021.
17 Buckingham narrates "the only two European consulships at Bagdad, are those of the English and French". See Buckingham, *Travels in Mesopotamia*, p. 389.
18 Rich, 1936, F&C, *Narrative of a residence in Koordistan, and on the site of ancient Nineveh*, p. xvii.
19 Bond, 'Claudius Rich and Samuel Manesty'.
20 Buckingham, *Travels in Mesopotamia*, p. 390.

21 Bond used the word 'immunise' because "some later travellers remarked that some who had lived in the East for too long appeared corrupted in character with traits of 'Oriental' manners. Samuel Manesty for one became a target of such mockery". Bond, 2016, 'Claudius Rich and Samuel Manesty'.
22 Bond, 'Claudius Rich and Samuel Manesty'.
23 Bond, 'Claudius Rich and Samuel Manesty'.
24 Bond, 'Claudius Rich and Samuel Manesty'.
25 Alexander, *Baghdad in bygone days*, p. 48.
26 Heude, *A voyage up the Persian Gulf*, p. 103.
27 Buckingham, *Travels in Mesopotamia*, p. 390.
28 Simpson, 'Arab and Islamic culture and connections', pp. 16–18.
29 Buckingham, *Travels in Mesopotamia*, p. 398.
30 Buckingham, *Travels in Mesopotamia*, p. 399.
31 Buckingham, *Travels in Mesopotamia*, p. 501.
32 Buckingham, *Travels in Mesopotamia*, p. 499.
33 Woodward, M 2019, 'Claudius James Rich: administrator, traveller, author, and collector of manuscripts and antiquities' <https://qdl.qa/en/claudius-james-rich-administrator-traveller-author-and-collector-manuscripts-and-antiquities> viewed 1 September 2021.
34 Woodward, 'Claudius James Rich: administrator, traveller, author, and collector of manuscripts and antiquities'.
35 See Alexander, *Baghdad in bygone days*. Also see Rich, *Narrative of a residence in Koordistan, and on the site of ancient Nineveh*.
36 Alexander, *Baghdad in bygone days*, p. 30.
37 Alexander, *Baghdad in bygone days*, pp. 30, 31.
38 See al-Attar, *Baghdad: an urban history through the lens of literature*, pp. 51–94.
39 An English man who had the position of high court judge of Bombay in India.
40 Buckingham, *Travels in Mesopotamia*, p. 392.
41 Alexander, *Baghdad in bygone days*, p. 33.
42 Jones, *Memoirs of Baghdad, Kurdistan and Turkish Arabia*.
43 Warren & Fethi, *Traditional houses in Baghdad*, p. 18.
44 Al-Warrak, *Baghdad bi'aqlām raḥḥāla*, p. 12.
45 Bianca, S & Eidgenossische Technische Hochschule Zurich. Institut fur Orts-Regional- und Landesplanung 2000, *Urban form in the Arab world: past and present*, VDF, Zurich, p. 66.
46 Alexander, *Baghdad in bygone days*, p. 35.
47 Alexander, *Baghdad in bygone days*, p. 34.
48 Narrow lanes in the neighbourhoods were meant to cool the alleys.
49 Heude, *A voyage up the Persian Gulf and*, p. 96.
50 Al-Warrak, *Baghdad bi'aqlām raḥḥāla*, p. 373.
51 Al-Warrak, *Baghdad bi'aqlām raḥḥāla*, p. 102.
52 Al-Warrak, *Baghdad bi'aqlām raḥḥāla*, p. 102.
53 For example, in regard to some mosques in Baghdad, he notes: "the whole is mush inferior to the Turkish minarets of Syria, and still more so to the light and elegant ones seen in many parts of Egypt". See Buckingham, *Travels in Mesopotamia*, p. 375.
54 Buckingham, *Travels in Mesopotamia*, p. 367.
55 Buckingham, *Travels in Mesopotamia*, p. 374.
56 Buckingham, *Travels in Mesopotamia*, p. 374.

57 Coke, *Baghdad; the city of peace*, p. 239.
58 Buckingham, *Travels in Mesopotamia*, pp. 370, 371.
59 Coke, *Baghdad; the city of peace*, p. 240.
60 Buckingham, *Travels in Mesopotamia*, p. 371.
61 Buckingham, *Travels in Mesopotamia*, p. 380.
62 This word was written in Buckingham's book like this. Elsewhere in this book I transliterated it as *sirdāb* to designate basement rooms in the courtyard houses.
63 Buckingham, *Travels in Mesopotamia*, p. 380.
64 Buckingham, *Travels in Mesopotamia*, p. 380.
65 Buckingham, *Travels in Mesopotamia*, p. 372.
66 The western part of Baghdad, or *al-karkh*, has always been inhabited through different stages of the city's history, and it has been one of the key components of the city, yet it was smaller than the eastern side.
67 Buckingham, *Travels in Mesopotamia*, p. 395.
68 Issawi, CP 1980, *The economic history of Turkey, 1800–1914*, University of Chicago Press, Chicago, IL, p. 17.
69 Al-Warrak, *Baghdad bi'aqlām raḥḥāla*, p. 105.
70 Buckingham, *Travels in Mesopotamia*, p. 372.
71 Buckingham, *Travels in Mesopotamia*, p. 372.
72 Buckingham, *Travels in Mesopotamia*, p. 383.
73 Al-Warrak, *Baghdad bi'aqlām raḥḥāla*, pp. 103, 109.
74 See al-Warrak, *Baghdad bi'aqlām raḥḥāla*, p. 104. Also see Coke, *Baghdad; the city of peace*, p. 241.
75 Al-Warrak, *Baghdad bi'aqlām raḥḥāla*, pp. 103, 104.
76 Coke, *Baghdad; the city of peace*, p. 241.
77 Buckingham, *Travels in Mesopotamia*, p. 377.
78 Al-Warrak, *Baghdad bi'aqlām raḥḥāla*, pp. 137, 138.
79 Buckingham, *Travels in Mesopotamia*, p. 394.
80 Al-Warrak, *Baghdad bi'aqlām raḥḥāla*, p. 138.
81 Buckingham, *Travels in Mesopotamia*, p. 515. Also see Al-Warrak, *Baghdad bi'aqlām raḥḥāla*, p. 137.
82 Buckingham, *Travels in Mesopotamia*, p. 514.
83 Buckingham, *Travels in Mesopotamia*, p. 373.
84 Buckingham, *Travels in Mesopotamia*, pp. 372–373.
85 Al-Warrak, *Baghdad bi'aqlām raḥḥāla*, p. 105.
86 Buckingham, *Travels in Mesopotamia*, p. 374.
87 Conventional history on Islamic cities often highlights the street layout in these cities in relation to the urban morphology of Islamic cities, and to issues of modernisation that imply wider streets.
88 See Buckingham, *Travels in Mesopotamia*, pp. 376, 379. Also see Al-Warrak, *Baghdad bi'aqlām raḥḥāla*, pp. 111, 112.
89 Buckingham, *Travels in Mesopotamia*, p. 379.
90 Buckingham, *Travels in Mesopotamia*, p. 376.
91 Buckingham, *Travels in Mesopotamia*, p. 376.
92 Buckingham, *Travels in Mesopotamia*, pp. 513, 514.
93 Buckingham, *Travels in Mesopotamia*, p. 513.
94 Buckingham, *Travels in Mesopotamia*, p. 513.
95 Buckingham, *Travels in Mesopotamia*, p. 379.
96 Buckingham, *Travels in Mesopotamia*, p. 378.

 97 Buckingham, *Travels in Mesopotamia*, p. 374.
 98 Buckingham, *Travels in Mesopotamia*, pp. 378, 512. These mosques still exist today.
 99 Buckingham, *Travels in Mesopotamia*, p. 513.
100 Buckingham, *Travels in Mesopotamia*, p. 377.
101 Buckingham, *Travels in Mesopotamia*, p. 377.
102 Buckingham, *Travels in Mesopotamia*, p. 375.
103 Buckingham, *Travels in Mesopotamia*, p. 376.
104 Buckingham, *Travels in Mesopotamia*, p. 377.
105 Buckingham, *Travels in Mesopotamia*, p. 378.
106 Al-Warrak, *Baghdad biʾaqlām raḥḥāla*, p. 110.
107 Buckingham, *Travels in Mesopotamia*, p. 378.
108 Examples of Baghdadi Scholars' writings in al-Attar, *Baghdad: an urban history through the lens of literature*.
109 Buckingham, *Travels in Mesopotamia*, p. 380.
110 Heude, *A voyage up the Persian Gulf*.
111 Coke, *Baghdad, the city of peace*, p. 239.
112 The Madras Army was the army of the Presidency of Madras, one of the three presidencies of British India within the British Empire.
113 Earl Fitzwilliam was a title in both the Peerage of Ireland and the Peerage of Great Britain held by the head of the Fitzwilliam family.
114 Heude, *A voyage up the Persian Gulf*, p. 177.
115 Heude, *A voyage up the Persian Gulf*, p. 187.
116 Heude, *A voyage up the Persian Gulf*, p. 187.
117 Heude, *A voyage up the Persian Gulf*, p. 187.
118 Heude, *A voyage up the Persian Gulf*, p. 138.
119 Heude, *A voyage up the Persian Gulf*, p. 187.
120 Heude, *A voyage up the Persian Gulf*, p. 92.
121 Heude, *A voyage up the Persian Gulf*, p. 92–93.
122 Heude, *A voyage up the Persian Gulf*, p. 139.
123 Heude, *A voyage up the Persian Gulf*, p. 159.
124 Heude, *A voyage up the Persian Gulf*, p. 182.
125 Heude, *A voyage up the Persian Gulf*, p. 115.
126 Simpson, 'Arab and Islamic culture and connections', pp. 16–18.
127 Heude, *A voyage up the Persian Gulf*, p. 107.
128 Fruitt, S 2016, 'Where did the word barbarian come from' <https://www.history.com/news/where-did-the-word-barbarian-come-from> viewed 10 September 2021.
129 Tuʿma, T 2021, 'man hum al-barbar' (who are the barbar?), <https://sotor.com>, viewed 10 September 2021.
130 Baheth <http://baheth.info> بربر, viewed 12 September 2021.
131 Fruitt, 'Where did the word barbarian come from'.
132 Fruitt, 'Where did the word barbarian come from'.
133 Heude, *A voyage up the Persian Gulf*, p. 166.
134 Heude, *A voyage up the Persian Gulf*, p. 120.
135 Heude, *A voyage up the Persian Gulf*, p. 121.
136 Heude, *A voyage up the Persian Gulf*, p. 126.
137 Heude, *A voyage up the Persian Gulf*, p. 127.
138 Heude, *A voyage up the Persian Gulf*, p. 125.
139 Fogg, *Arabistan, or the land of the Arabian nights*, p. 232.

140 Simpson, 'Arab and Islamic culture and connections', pp. 16–18.
141 Simpson, 'Arab and Islamic culture and connections', pp. 16–18.
142 Heude, *A voyage up the Persian Gulf*, p. 129.
143 Heude, *A voyage up the Persian Gulf*, p. 180.
144 Heude, *A voyage up the Persian Gulf*, pp. 129–130.
145 Heude, *A voyage up the Persian Gulf*, p. 180.
146 Heude, *A voyage up the Persian Gulf*, p. 180.
147 Heude, *A voyage up the Persian Gulf*, p. 180.
148 Heude, *A voyage up the Persian Gulf*, pp. 180, 181.
149 Heude, *A voyage up the Persian Gulf*, p. 180.
150 Heude, *A voyage up the Persian Gulf*, p. 181.
151 Heude, *A voyage up the Persian Gulf*, p. 181.

7 Summary

The analysis of the travelogues in this book established a more realistic vision of this kind of literature and revealed additional aspects that are not commonly observed or examined. The discussion of these texts designated their potential to disclose further aspects of the urban history of Baghdad and offered a 'particular spatialisation of knowledge',[1] along with other sources in historiography. It also presented a strong distinction between the explorative attitudes of regional travellers and the interfering approaches of the emissaries. Unlike regional explorers who were strongly engaged with discovery, learning and exploring, European emissaries' observations were part of their job to provide evidence of the diversity of people, places and manners that existed in the cities they visited, to assist them in future colonisation of these lands.

Generally, the techniques of European travel literature about Baghdad involve a detailed description of the city, its architecture, political situation, trade, history, geography and people. The reflection of each traveller depended on a number of factors, including the purpose of the visit, available information, personal education, possible access to places and historical texts, and the relevance of their writing to the interest of the reading public and publishers. Although the travelogues of the 18th and 19th centuries offered imprints of visual and geographical aspects, and encompassed descriptive narratives and topographical overviews, they reflected the traveller's own expressions rather than the city's real meaning. This book mainly examined the writings of six emissaries with diverse backgrounds. These texts show a strong connection between the objective of their visit and the interpretation of the city. While Niebuhr was primarily a surveyor, he expressed interest in the history of Baghdad, and he wrote about its urban forms. Similarly, Olivier was an entomologist, but he expanded his writings to encompass some architectural details, in addition to other historical accounts of Baghdad. The detailed map by Niebuhr, and

DOI: 10.4324/b23141-8

Olivier's ecological material, assist the reading of the history of Baghdad in the late decades of the 18th century.

Likewise, the emissaries of the early 19th century wrote comments about Baghdad in that period. Because Rich was involved in political matters, he expressed little information about the urban settings of Baghdad. Yet the letters of his wife, Mary, comprised more comments about the city's conditions. Likewise, Buckingham provided descriptions of Baghdad at the time of his visit. The fourth emissary of the 19th century, Heude, clearly unveiled his colonial intentions, outlining Baghdad as a city that "can best mark the character of despotic states, and the condition of the subject under the influence of an unlimited authority".[2] He provided strategies for invasion by the British. His intrusive attitudes paved the way for the conquest later, by continuously criticising the political situation. Despite slight differences between their backgrounds, all of these emissaries reflected parallel terminologies through their texts. Since colonisation was their main goal, it is reasonable to call this kind of literature 'colonial literature'.

One of the most destructive tricks of the colonial literature is exaggeration and invention. In general, many emissaries had extremely limited contact with the societies they visited. Yet, they wrote about those societies comprehensively, which resulted in a huge number of invented events. Buckingham asserts this situation stating, "for, like Herodotus, and indeed many more modern travellers, whenever he [Tudela] quits the boundaries of his own observation, all is fable and exaggeration".[3] An example of this misrepresentation is the overemphasis on the bad image of Baghdad by pointing out a gloomy visual appearance and an insecure political situation. Another problem is selectivity and lack of acknowledgement of information that does not serve their purposes, in addition to the lenience towards seeking and transmitting truthful information, which resulted in unjust statements. In fact, untruthful commentary by emissaries is common for other the Arab cities, in general, that "received what often seems to be more than its fair share of inaccurate reporting by Western visitors ... they brought with them a great potential for international misunderstanding".[4]

The strong attention to the physical appearance and considering it a main aspect of appreciation is another shortage of this kind of literature. Even when the emissaries depicted a positive atmosphere in Baghdad, they did not reflect on its social and spiritual beauty, which makes the picture incomplete. This approach delivered a partial understanding of Baghdad and brought up contradictory visions as to its meaning. For example, while colonial literature conveyed a negative image of Baghdad in the 18th and 19th centuries, this period produced valuable texts by Baghdadi scholars that draw a beautiful image, underlined by the social, spiritual and materialistic beauty of the city. Moreover, European travelogues lacked serious historical grounding. Instead of researching and analysing historical information deeply, the emissaries depended on imaginative old tales and on what they heard from people, considering that material among their resources for history. As a result, colonial literature provided a transient representation of the history of Baghdad, and a plain description that is emptied of any meaning beyond the writer's specific objectives. This insufficient method of historical inquiry emphasises the necessity to recognise historiographical sources before implementing them.

The emissaries' strong devotion to history writing later opened the door to much misinterpretation and invention. Lewis notes, "probably the outstanding example in our time of the inventive and purposive use of historiography is the writing of colonial, post-colonial, and finally pre-colonial history".[5] These attitudes made these travelogues a favourite body of literature that promoted the Orientalists' method of a plain appreciation of the 'Orient'. Subsequently, they facilitated colonialism, which brought terrible effects to many countries, including Iraq, lasting until the present day. European emissaries were specifically attached to the glorious pictures of the partially imaginative tales of the *Arabian nights* that were composed about ten centuries prior to their visits to Baghdad. The extreme reliance on these tales points out the power of fiction in influencing literature. Above all, European travelogues promoted the strategy of 'stereotyping' to deal with people who are different from Westerners. This method of differentiation that is consistent with specific physical differences and norms, could be considered the most damaging product of the colonial literature. This concept implements ideas such as backwardness, accusation of others, humiliating dark-skinned people and weakening the norms of other societies. Unfortunately, these ideas continue to be effective today, as "progress so far has been very mixed, particularly in overcoming stereotypes of Arabs that are rooted in Western attitudes to the Middle East".[6] As a result, these tactics brought harmful consequences to many lands in the Arab world.

In short, this study has proved that colonial literature is not a typical travel literature, and thus historians cannot rely on it completely to obtain facts. The capacity of history should not be limited to monotonous views that are embedded in European travelogues, for example. Rather, history should consider all other sources, and examine the circumstances that influence the writing procedure. In addition, the method of representation should not be restricted to certain aspects that eliminate the conditions of history writing. To evade misunderstanding, it is crucial to study all periods in its long history, rather than focussing on a single period in history. The extensive emphasis of European travelogues on dates, places, and other statistics makes them a rich source of information for transitional histories. However, they should be observed in conjunction with other resources, to reveal additional meanings through comparison and deep analyses of different periods and various circumstances, to recognise their interconnectivity. Furthermore, the historical information provided by European travelogues should be examined carefully. Along with any other historical reference, they cannot be considered completely truthful. These steps would enhance the implication of history in general, and the understanding of the history of Baghdad in particular. In order to discontinue its damaging effects, the negative ideas that are initiated by the colonial literature should be addressed and resolved, including stereotyping, backwardness, superiority and all other Orientalist models. In this way, we can obtain the most possible benefit from all human experiences, including travelogues.

Notes

1 Boyer, *The city of collective memory,* p. 133.
2 Heude, *A voyage up the Persian Gulf,* p. 177.
3 Buckingham, *Travels in Mesopotamia,* p. 504.
4 Simpson, 'Arab and Islamic culture and connections', pp. 16–18.
5 Lewis, B 1975, *History: remembered, recovered, invented,* Princeton University Press, Princeton, NJ, p. 87.
6 Simpson, 'Orientalist Arab and Islamic culture and connections', pp. 16–18.

Bibliography

English references

Abdullah, T 2001, *Merchants, Mamluks, and murder: the political economy of trade in eighteenth-century Basra*, SUNY series in the social and economic history of the Middle East, State University of New York Press, Albany, Britain.

Aga Khan Award for Architecture 1986, *Architecture education in the Islamic world: proceedings of seminar ten*, Granada, Spain, Concept Media Pte. Ltd., Singapore.

Akbar, JA 1984, 'Responsibility in the traditional Muslim built environment', PhD thesis, Massachusetts Institute of Technology.

Akkach, S 2002, 'On culture', in Akkach, S & University of Adelaide, Centre for Asian & Middle Eastern Architecture (eds.), *De-placing difference: architecture, culture and imaginative geography*, Centre for Asian and Middle Eastern Architecture, the University of Adelaide, pp. 183–189.

Al-Attar, I 2018, *Baghdad: an urban history through the lens of literature*, Routledge, London and New York.

Alexander, CM 1928, *Baghdad in bygone days: from the journals and correspondence of Claudius Rich, traveller, artist, linguist, antiquary and British resident at Baghdad, 1808–1821*, 1st edn, J. Murray, London.

Ansari, MT (ed.) 2002, *Secularism, Islam, modernity: selected essays of 'Alam Khundmiri*, SAGE, London.

Atasoy, N 2004, 'Ottoman garden pavilions and tents', in *Muqarnas*, vol. 21, *Essays in honor of J. M. Rogers*, the Agha Khan Program for Islamic Architecture at Harvard University and the Massachussets Institute of Technology, Cambridge, Massachusetts, pp. 15–19.

Ayduz, S 2008, 'Nasuh Al-Matraki: a noteworthy Ottoman artist-mathematician of the sixteenth century', viewed 5/1/2014, <MuslimHeritage.com>.

Bacon, EN 1967, *Design of cities*, Thames and Hudson, London.

Bianca, S & Eidgenossische Technische Hochschule Zurich. Institut fur Orts-Regional- und Landesplanung 2000, *Urban form in the Arab world: past and present*, VDF, Zurich.

Black, E 2004, *Banking on Baghdad*, John Wiley & Sons, Inc., Hoboken, NJ.

Bond, XW 2016, 'Claudius Rich and Samuel' Manesty' <https://blogs.bl.uk/untoldlives/2016/03/claudius-rich-and-samuel-manesty.html> viewed 3/9/2021.

Bonine, ME 2005, 'Islamic urbanism, urbanites, and the Middle Eastern city', in Choueiri, YM (ed.), *A companion to the history of the Middle East*, Blackwell companions to world history, Blackwell Pub. Ltd, Malden, MA, pp. 393–406.

Boyer, MC 1994, *The city of collective memory: its historical imagery and architectural entertainments*, MIT Press, Cambridge, MA.

Buckingham, JS 1827, *Travels in Mesopotamia, including a journey from Aleppo, across the Eurphrates to Orfah, (the Ur of the Chaldees) through the plains of the Turcomans, to Diarbeker, in Asia Minor; from thence to Mardin, on the borders of the Great Desert, and by the Tigris to Mousul and Bagdad; with researches on the ruins of Babylon, Nineveh, Arbela, Ctesiphon, and Seleucia*, Henry Colburn, London.

Cerasi, M 2005, 'The urban and architectural evolution of the Istanbul divanyolu: urban aesthetics and ideology in Ottoman town building', *Muqarnas*, vol. 22, pp. 189–232.

Chard, C 1999, *Pleasure and guilt on the grand tour: travel writing and imaginative geography, 1600–1830*, Manchester University Press, Manchester.

Chisholm, H 1911, *The Encyclopaedia Britannica*, vol. 7, Constantinople, the capital of the Turkish Empire, Internet Archive HTML5 Uploader 1.6.4.

Coke, R 1927, *Baghdad: the city of peace*, Thornton Butterworth LTD, London, p. 239.

Cooperson, M 1996, 'Baghdad in rhetoric and narrative', in *Muqarnas: an annual on the visual culture of the Islamic world. Volume 13, Aga Khan Program for Islamic architecture*, Cambridge, pp. 99–113.

Choueiri, YM 2005, *A companion to the history of the Middle East*, Blackwell companions to world history, Blackwell Pub. Ltd, Malden, MA.

Dabrowska, K & Hann, G 2008, *Iraq then and now: a guide to the country and its people*, Bradt Travel Guides, Chalfont St. Peter, p. 199.

Encyclopaedia Britannica, <https://en.wikisource.org> viewed 10/03/2020.

Ettinghausen, R, Grabar, O & Jenkins, M 2001, *Islamic art and architecture 650–1250*, Yale University Press Pelican history of art, Yale University Press, New Haven, CT.

Fogg, WP 1985, *Arabistan, or the land of the Arabian nights: being travels through Egypt, Arabia, and Persia, to Bagdad*, Darf, London.

Fruitt, S 2016, 'Where did the word barbarian come from'? <https://www.history.com/news/where-did-the-word-barbarian-come-from> viewed 10/9/2021.

Grabar, O 1987, *The formation of Islamic art*, Rev. and enl. edn, Yale University Press, New Haven, CT.

Hamadeh, SH 2007, 'Public spaces and the garden culture of Istanbul in the eighteenth century', in Aksan, VH & Goffman, D (eds.), *The early modern Ottomans: remapping the Empire*, Cambridge University Press, Cambridge, pp. 299–310.

Hopkins, IWJ 1967, 'The maps of Carsten Niebuhr: 200 years after', *Cartographic Journal*, vol. 4, no. 2, pp. 115–118.

Heude, W 1970, *A voyage up the Persian Gulf and a journey overland from India to England in 1817*, containing Notices of Arabia Felix, Arabia deserta, Persia ... Bagdad, Koordistan, Armenia, Asia Minor, London, Knightsbridge.

Inalcik, H & Quataert, D 1997, *An economic and social history of the Ottoman Empire, Volume 2: 1600–1914*, 2 vols, Cambridge University Press, Cambridge.

Issawi, CP 1980, *The economic history of Turkey, 1800–1914*, University of Chicago Press, Chicago, IL, p. 17.

Jones, JF 1998, M*emoirs of Baghdad, Kurdistan and Turkish Arabia*, 1857, selections from the records of the Bombay Government, number xliii, new series, Archive Editions, Great Britain.

Kafesci oglu, C 2009, *Constantinopolis/Istanbul: cultural encounter, imperial vision, and the construction of the Ottoman capital*, The Pennsylvania State University Press, University Park, PA.

Khoury, DR 2007, 'Who is a true Muslim? Exclusion and inclusion among polemicists of reform in nineteenth-century Baghdad', in Aksan, VH & Goffman, D (eds.), *The early modern Ottomans*, Cambridge University Press, Cambridge, UK, pp 256–274.

Lafi, N 2007, 'The Ottoman municipal reforms between old regime and modernity: Towards a new interpretive paradigm', in Cihangir, E (ed.), *Uluslararas Eminonu Sempozyumu: tebligler kitab International Symposium on Eminonu: the book of notifications*, Eminonu Belediyesi Baskanlg, Istanbul, pp. 348–358.

Lassner, J 1970, *The topography of Baghdad in the early middle ages: text and studies*, Wayne State University Press, Detroit.

Lewis, B 1975, *History: remembered, recovered, invented*, Princeton University Press, Princeton, NJ.

Longrigg, SH 1968, *Four centuries of modern Iraq*, Librairie du Liban, Beirut.

Moholy-Nagy, S 1968, *Matrix of man: an illustrated history of urban environment*, Pall Mall Press, London.

Morkoc, SB 2010, *A study of Ottoman narratives on architecture: text, context and hermeneutics*, Academia Press, Bethesda Cture.

Nasar, J 1998, *The evaluative image of the city*, Sage Publications, Thousand Oaks, CA.

Niebuhr, C & Heron, R 1792, *Travels through Arabia and other countries in the East*, R. Morison and Son, Edinburgh.

Niebuhr, C (1733–1815) 1983, *Entdeckungen im Orient: Reise nach Arabien und anderen Ländern 1761–1767*, K. Thienemanns Verlag, Stuttgart.

Oktay, B, Elwazani, S, & al-Qawasmi, J (eds.) 2008, *Responsibilities and opportunities in architectural conservation; theory, education, & practice*, vol. 2, CSAAR, Amman.

Ousterhout, R, Necipoglu Gl & Aga Khan Program for Islamic Architecture (eds.) 1995, 'Ethnic identity and cultural appropriation in early Ottoman architecture', in *Muqarnas: an annual on Islamic art and architecture*, vol. 12, E.J. Brill, Leiden, the Netherlands, pp. 48–62.

Rapoport, A 1977, *Human aspects of urban form: towards a man-environment approach to urban form and design*, 1st edn, Urban and regional planning series, v. 15, Pergamon Press, Oxford.

Raymond, A 2002, *Arab cities in the Ottoman period: Cairo, Syria, and the Maghreb*, Ashgate, Variorum, Aldershot, Hampshire, Great Britain; Burlington, VT.

Rich, CJ 1836, *Narrative of a residence in Koordistan, and on the site of ancient Nineveh: with journal of a voyage down the Tigris to Bagdad and an account of a visit to Shirauz and Persepolis*, 2 vols., J. Duncan, London.

Said, EW 1979, *Orientalism*, 25th anniversary edn, Vintage Books, New York, p. 47.

———— 1985, *Beginnings: intention and method*, Columbia University Press, New York.

———— 2007, O*n late style: music and literature against the grain*, 1st Vintage Books edn, Vintage Books, New York.

Scoville, S 1977, 'Beshreibung von Arabian by Carsten Niebuhr' (book review), *International Journal of Middle East Studies*, vol. 8, no. 2, pp. 275–276.

Simpson, M 1989, 'Orientalist travellers, Aramco World; Arab and Islamic culture and connections', vol. 40, no. 4, pp. 16–18. <https://archive.aramcoworld.com/issue/198904/Orientalist.travelers.htm> viewed 4/3/2020.

Sinclair, WF 1967, *The travels of Pedro Teixeira, with his 'Kings of Harmuz' and extracts from his 'Kings of Persia'*, with further notes and an introduction by Donald Ferguson, Hakluyt Society, Kraus Reprint, Nendeln, Liechtenstein.

Society for the Diffusion of Useful Knowledge 1833, *Lives of eminent persons*, Baldwin and Cradock, London.

Soder, H 2003, 'The return of cultural history; literary historiography from Nietzsche to Hayden White', H*istory of European Ideas*, vol. 29, no. 1, pp. 73–84.

Stark, F 1947, *Baghdad sketches*, published for Guild by J. Murray, London.

Tavernier, JB, Crooke, W & Ball, V 1977, T*ravels in India*, 2nd edn, 2 vols, Oriental Books Reprint Corporation, New Delhi.

Tripp, C 2007, *A history of Iraq*, 3rd edn, Cambridge University Press, Cambridge.

Veinstein, G 2008, 'The Ottoman town; fifteenth-eighteenth centuries', in Jayyusi, SK, Holod, R, Petruccioli, A & Raymond, A (eds.), *The city in the Islamic world*, Brill, Leiden, and Boston, MA, pp.207–212.

Vernoit, SJ 2007–2012, 'Niebuhr, Carsten', in Oxford Art Online, Oxford University Press, <http://www.oxfordartonline.com:80/subscriber/article/grove/art/T062405>, viewed 4/5/2013.

Warren, J & Fethi, I 1982, *Traditional houses in Baghdad*, Coach Publishing House, Horsham, p. 18.

Wellsted, JR & Ormsby 1968, *Travels to the city of the Caliphs, along the shores of the Persian Gulf and the Mediterranean: including a voyage to the coast of Arabia, and a tour on the Island of Socotra*, 2 vols, Gregg International, Farnborough, Hants.

Wilson, D (ed.) 1976, *Islam and medieval Hellenism: social and cultural perspectives*, Variorum reprints, London.

Woodward, M 2019, 'Claudius James Rich: administrator, traveller, author, and collector of manuscripts and antiquities' <https://qdl.qa/en/claudius-james-rich-administrator-traveller-author-and-collector-manuscripts-and-antiquities> viewed 1/9/2021.

Arabic references

Al-Sadr, MB 2003, *al-Islam yaqūd al-hayāt, al-Madrasa al-Islāmiyya, risālatunā* (Islamic teachings guide life), Centre of special studies of Imam al-Sadr writings, Shariat, Qum.

Al-Warid, BA 1980, *ḥ 8ādith Baghdad fi 12 qarn* (The events of Baghdad through 12 centuries), al-dār al-ʿarabiyya, Baghdad.

Al-Warrak 2007, *Baghdad biʾaqlām raḥḥāla* (Baghdad in the writings of travellers), al-Warrak Publishing Ltd, London.

Baheth <http://baheth.info> بحث.

Faʾiq, S 2010, *tārīkh Baghdad* (The history of Baghdad), 1st edn, dār al-rāfidayn for publishing, Beirut.

Izz al-Din, Y 1976, *Dawud pāshā wa nihāyat al-mamālik fi al-Iraq* (Dawud pasha and the end of Mamluk's ruling in Iraq), ma udʿat al-shaʿab, the University of Baghdad, Baghdad.

Jawad, M & Susa, A 1958, *dalil khāriṭat Baghdad al-mufaexample khiṭaṭ Baghdad qadiman wa ample.co* (A guide to the map of Baghdad), al-majmaʿ al-ʿilmi al-Iraqi, Baghdad.

Jawad, M, Susa, A, Makkiya, M & Maʿruf, N 1968, *Baghdad*, Iraqi Engineers Association with Gulbenkian Foundation, Baghdad.

Khayyat, J (ed.) 1968, *ʾarbaʿat qurūn min tārīkh al-Iraq al-hadīth* (Four centuries of modern Iraq, by Stephen Hemsley Longrigg), matbaʿat al-tafayyud, Baghdad.

Khuja, KA 2006, 'muqtaṭafāt min kitāb gulshan khulafaʾ by Murtaḍa N, *Arabic Translators International* (parts of the book gulshan khulafaʾ), <http://www.atinternational.org/forums/showthread.php?t=7467>, viewed 4/8/2012.

Khulusi, S (ed.) 1962, *tārīkh Baghdad:hadiqat al-Zawrāʾ fī sīrat al-wuzarāʾ by ʿabd al-Rahman al-Suwaidi* (The history of Baghdad), vol. 1, mavolʿat al-Zaʿim, Baghdad.

Makkiyya, M (ed.) 2005, *Baghdad*, 1st edn, al-Warrak Publishing Ltd, London.

Mohammed Ali, IM 2008, *madinat Baghdad: al-ʿabʿād al-ijtimāʿiyya wa ẓurūf al-nashʾa* (Social aspects of Baghdad and its beginnings), al-haial aspects ṭibāʿa wal nashr, al-ʿārif lil maṭbūʿāt, Baghdad.

Nawras, AMK 1975, *ḥukkām al-mamālik 1750–1831* (Mamluck governors), silsilat al-kutub al-0-1831om?id=misc"-ʿārif lil maṭbūʿāth ational.org/forums/s number 611, Baghdad.

Raʾuf, IA (ed.) 1978, *tārīkh ḥawādith Baghdad wal-Basrah 1186–1192 AH, 1772–1778 AD* (History of the events of Baghdad and Basra), *by* ʿabd al-Rahmān al-Suwaidi, The Ministry of Education and Arts, Baghdad.

——— 2000, *maʿālim Baghdad fī al-qurūn al-mutaʾkhira* (The attributes of Baghdad in late centuries)), Bayt al-butea, Baghdad, Iraq.

———— 2004, *al-ʿiqd al-lāmiʿ bi-ʾāthār Baghdad wal-masājid wal-jawāmiʿ* (The heritage of Baghdad's mosques), by *'y heritage of Baghd*, 1st edn, ' nst itage of Baghding, Baghdad.

———— 2008, *' 08-ār Baghdad wa mā jāwarahā min al-bilād, by al-Alusi M Sh* (The events of Baghdad and the surrounding area), al-dār al-arabiyya lil mawsūʿāt, Beirut.

Selman, I, ʿ lmanINK "htt, H, al-Izzi, N, & Yunus, N 1982, *al-ʿimārāt al-ʿarabiyya al-Islāmiyya fī al-Iraq* (Arabic and Islamic buildings in Iraq), 1&2, al-ḥurriyya Press, Baghdad.

Susa, A 1952, *Atlas Baghdad*, mudiriyyat al-masāexampleāmma, Baghdad.

The Holy Qurʾan, <http://www.quranexplorer.com>, viewed 26/8/2021.

Tuʿma, T 2021, 'Man hum al-barbar' (Who are the barbar), <https://sotor.com>, viewed 10/9/2021.

Index

Printed and bound by CPI Group (UK) Ltd, Croydon, CR0 4YY

17/10/2024

01775689-0011